Daniel G. Elliot

A Monograph of the Tetraoninae

Family of the Grouse

Daniel G. Elliot

A Monograph of the Tetraoninae
Family of the Grouse

ISBN/EAN: 9783337330750

Printed in Europe, USA, Canada, Australia, Japan

Cover: Foto ©berggeist007 / pixelio.de

More available books at **www.hansebooks.com**

A

MONOGRAPH OF THE TETRAONINAE,

OR

FAMILY OF THE GROUSE.

BY

DANIEL GIRAUD ELLIOT,

FELLOW OF THE ZOOLOGICAL SOCIETY OF LONDON: OF THE NATIONAL ACADEMY OF DESIGN; MEMBRE DE LA SOCIÉTÉ ZOOLOGIQUE IMPÉRIALE
D'ACCLIMATION; MEMBER OF THE ACADEMY OF NATURAL SCIENCES OF PHILADELPHIA; OF THE NEW YORK LYCEUM OF
NATURAL HISTORY: OF THE NEW YORK HISTORICAL SOCIETY: CORRESPONDING MEMBER OF THE
BOSTON NATURAL HISTORY SOCIETY, &c., &c.

NEW YORK:
PUBLISHED BY THE AUTHOR, NO. 27 WEST THIRTY-THIRD STREET.
1865.

Professor Spencer F. Baird,

OF THE

SMITHSONIAN INSTITUTION, WASHINGTON,

THE RESULT OF WHOSE LABORS IN THIS FAVORITE BRANCH OF SCIENCE HAS DECISIVELY PLACED HIM AMONG
THE FIRST OF LIVING ORNITHOLOGISTS;

I DEDICATE THIS

MONOGRAPH OF THE TETRAONINAE,

OR

FAMILY OF THE GROUSE,

AS A TESTIMONY OF

FRIENDSHIP AND ESTEEM.

D. G. ELLIOT.

LIST OF SUBSCRIBERS.

ASHER, Messrs. & Co. London.
ACADEMY OF NATURAL SCIENCES, The. Philadelphia, per Dr. T. B. Wilson.

BANCROFT, Messrs. H. H. & Co. San Francisco, Cal. (Two Copies.)
BOERNOST, J. CARSON, Esq. Brooklyn, New York.
DRAKE, H.,,Esq. New York.
BROWN, G. H., Esq. New York.
BROCKHAUS, Mon. F. A. Leipsic.
BRITISH MUSEUM. The Library of the. London.

CHADWICK, J., Esq. New York.
CURRIER, JOHN MCNABB, M.D. Newport, Vermont.
CLARKE, ROBERT, Esq. Cincinnati, Ohio.
CLARKE, E. W., Esq. Philadelphia.
CLASSENIAN LIBRARY, The. Copenhagen.

DENNY, JOHN T. Esq. New York.
DEVILLISS, W. H., M.D. New York.

EARLE, JAMES, Esq. New York.
ELLIOT, A. F., Esq. New Orleans.
ELLIOT, G. T., Esq. New York.
EYTON, T. C., Esq. Eyton. Wellington, Shropshire, England.

FOSTER, F. G., Esq. New York.
FOSTER, H. T. E., Esq. Lakeland. Seneca Co., New York.
FOSTER, J. P. GIRAUD, Esq. New York.
FOWLER, MORTIMER L., Esq. New York.

GARRISON, J. H. Esq. Jersey City, New Jersey.
GOULD, JOHN, Esq. Author of the Birds of Europe; Birds of Australia;
   Birds of Asia; Birds of Great Britain, &c. London.
GRAHAM, J. LORIMER, Jr., Esq. New York.
GRINNELL, MOSES H., Esq. New York.

HAYS, JACOB, Esq. New York.
HAYS, W. J., Esq. New York.
HEPBURN, JAMES, Esq. San Francisco, Cal.
HERRICK, J. K., Esq. New York.
HOTCHKISS, H. L., Esq. New Haven, Conn.

LEDDON, FRANK, Esq. New York.

KIRTLAND, Prof. J. P., M.D., LL.D. Cleveland, Ohio.
KENNOTT, NATHANIEL, Esq. New York.

LENOX, JAMES, Esq. New York.
LAWRENCE, G. N., Esq. Member of New York Lyceum of Natural History;
   Corresponding Member of the Zoological Society of London; of the
   Academy of Natural Sciences of Philadelphia, &c., &c.

MARSHALL, CHARLES H., Jr., Esq. New York.
McCORMICK, C. H., Esq. Chicago, Ill.
MILNE-EDWARDS, Mon. le Prof. Paris.
MUSEUM OF NATURAL HISTORY. The Library of. Paris.
MORRIS, J. B., Esq. New York.

NEWBOLD, GEORGE, Esq. New York.
NORRIS, J. P., Esq. Philadelphia, Pa.

OSGOOD, FRANKLIN, Esq. New York.

PALMER, R. S., Esq. New York.
PENDLETON, W. H., Esq. New York.

ROYAL LIBRARY OF BERLIN, The. Berlin.
ROYAL LIBRARY OF MUNICH, The. Munich.
ROGERS, GEORGE A., Esq. New York.
RUCKER, JOHN, Esq. New York.

SATTERLEE, LIVINGSTON, Esq. New York.
SMITHSONIAN INSTITUTE, The. Washington, D. C.
SMITH, G. G., Esq. New York.
STATE LIBRARY, The. Albany, New York.
STIMSON, A. E., Esq. Albany, New York. (Three Copies.)
STUART, R. L., Esq. New York. (Two Copies.)

THOMPSON, DANIEL, Esq. Chicago, Ill.
TRÜBNER, Messrs. & Co. London. (Three Copies.)
TURATI, Mon. le COMTE ERCOLE. Milan.

WILD, ALFRED, Esq. Albany, New York.
WILSON, Dr. T. B. Philadelphia.
WOOLSEY, W. W., Esq. New York.

YALE COLLEGE, The Library of. New Haven.

ZOOLOGICAL SOCIETY, The. London.

PREFACE.

In choosing a family of birds for a second Monograph, I was induced to make a selection of the Tetraonidæ, not only on account of their varied forms and interesting habits, but also for the important part they bear toward man's comfort and happiness. Although not brilliant in plumage of varied colors, like the Pittidæ, still few could witness the graceful forms, erect carriage, and gallant bearing of the members of this family, without having their admiration excited. The majority of the species are inhabitants of North America, and many of them, through the continued persecutions to which they are subjected, and the want of a rigid enforcement of proper laws for their protection, are rapidly disappearing from our land, in a comparatively short space of time to become extinct; and this was an additional reason to write their history while they were still to be found, and their habits observed in their native wilds.

Induenced by such motives, it was with no ordinary degree of interest that I entered upon my task, and have now brought the work to a conclusion, embracing within it all the species of the Tetraonidæ known to Ornithologists at the present time. "What is writ, is writ; would it were worthier."

In treating of so difficult a group as the Lagopidæ, or Ptarmigan, it was absolutely necessary, in order to arrive at just conclusions regarding the identification of the species, that a large number of specimens should be obtained; and I was particularly fortunate in receiving the vast collections of these birds made by Mr. Kennicott, during a protracted sojourn in Arctic America, as well as, from various other sources, numerous examples from almost every locality where these birds were known to exist; forming altogether probably the largest and most complete collection ever brought together. Therefore, after much investigation and study, it is with some degree of confidence that I have designated what have seemed to be good species; and although my fellow-Ornithologists may not agree with me in some of my views, yet from the material in my possession it was impossible for me to come to any other conclusions than those given in the various articles on this group; and in several instances record as synonyms local forms regarded by many as good species.

And now the pleasing duty devolves upon me, of acknowledging the assistance I have received in the prosecution of this work from my friends both in this country and in Europe: and first to Prof. S. F. Baird, of the Smithsonian Institution, Washington, who placed at my disposal all the material relating to this subject, gathered by the various collectors of the above Institution, and who has at all times given every aid in his power toward the successful completion of this Monograph; to Mr. John Cassin, of Philadelphia. I desire to express my obligations for advice regarding the preparation of my plates, and assistance at different periods cheerfully rendered; to Mr. Geo. N. Lawrence, who allowed me to appropriate for my use any specimens of these birds which his cabinet contained; to Mr. J. D. Sargeant, of Philadelphia, from whose fine examples of Canace Canadensis my drawing was made; to Mr. Alfred Newton, who sent me a fine series of Ptarmigan from Iceland, which were most useful in my investigations. To Dr. P. L. Sclater; Prof. Sundevall, of Stockholm; W. J. Hays, Esq.; Geo. A. Boardman, Esq.; Mons. Jules P. Verreaux; Benj. Leadbeater, Esq.; J. G. Bell, Esq.; John Krider, Esq., and others. I would here express my thanks for the aid given at various times. To John Gould, Esq., who sent from London many species of this family, together with his type of the Spitzbergen Ptarmigan, for my inspection, I am much indebted.

The plates furnished by Messrs. Bowen & Co., of Philadelphia, have been prepared with the usual care of that well-known firm; so long celebrated for their skilful execution in this difficult and delicate branch of art.

To Mr. C. F. Tholey, I would here state my gratification at the careful manner in which he has lithographed my drawings.

And now, nothing remains save to express the obligations I feel to those who have honored my work with their support, and with much patience have borne with its necessarily slow issue; whose assistance has encouraged me throughout my labors, and been the means of enabling me to bring them to a successful termination.

INTRODUCTION.

The Order Rasores, or Scrapers, so called from the habit possessed by its members of scratching the ground for the purpose of procuring their food, contains the most important species, for man, of all those included in the class Aves. It has its representatives in nearly every portion of the world, and comprises some of the most gorgeously-plumaged birds known to Ornithologists.

In Asia, probably the most typical groups and the greatest variety of forms occur. There the stately Peacock finds its natural home, and roams about in flocks of hundreds of individuals; while upon the mountains and in the forests many species of Pheasants dwell.

Among these last, distinguished for their beauty, I may here enumerate the Argus Giganteus or Argus Pheasant, remarkable for the extraordinary length of the secondary feathers, covered, as is the rest of its plumage, with numerous ocellated spots or eyes; the Lophophorus Impeyanus, or Monaul, whose bright metallic hues rival those of the humming-bird in their over-changing beauty; the Thaumalea Picta, or Golden Pheasant, with its splendid ruff of gold bordered with velvety black, its deep-red breast, and long, tapering tail-feathers; while many others with equal claims for an "honourable mention" might be named.

The Order consists of six families: the Cracidæ, or Curassows, nearest allied to the Columbidæ, or Pigeons, large birds, some species almost rivalling the Turkey in size, chiefly arboreal in their habits, and are inhabitants of South America; the Megapodidæ, or Mound Birds, a very extraordinary group, peculiar to Australia and the Malayan Archipelago, noted for laying large eggs and depositing them beneath piles of decaying vegetable matter, where they are hatched by the heat of the accumulated mass; the Phasianidæ, or Pheasants; the Tetraonidæ, or Grouse and Partridges; the Tinamidæ, or Tinamous, natives of South America; and the Pteroclidæ, or Sand Grouse. The Chionidæ, included with the above families by Gray and Bonaparte, should be omitted, as they are plavaline and not gallinaceous birds; the Chionis Alba approaching very closely in its osteological structure to Hæmatopus Niger.

The Tinamous may also, on account of their struthious characters, with some degree of propriety be separated from the gallinaceous birds; and although the Sand Grouse resemble in certain particulars both the Grouse and Pigeons, yet they belong to neither of these, and may be placed between the Grouse and Tinamous, these last tending to the Struthionidæ.

The families of this Order have been divided into many sub-families composed of numerous genera, and the one to which it is necessary for us now to turn our attention is that of the Tetraonidæ, which comprises the birds forming this Monograph. Many of the species are polygamous, the hens generally very prolific, gregarious in their habits, more or less capable of domestication; and, as they never wash, are accustomed to cleanse their feathers by rolling in the dust.

The Grouse are confined to the northern portions of Europe, Asia, and North America, and are rarely found in the warmer parts of those countries; while the Lagopidæ, or Ptarmigan, which constitute an important part of the family, are well called "children of the snow," and inhabit the high latitudes of both continents, having been discovered dwelling on the borders of the Arctic Sea. None of these have ever been found in Asia, although the lofty summits of the Himalayas would seem to be their natural abode; but their places are supplied in these regions by the splendid species of the genus Tetraogallus, or Snow-Partridges, which live upon the lofty heights of the mountains, and only in summer descend to the borders of vegetation. As yet no Ptarmigan have been discovered in Africa, where, upon the Mountains of the Moon, they might be supposed to exist, nor in any portion of South America, although the lofty ranges of the Andes would afford them a congenial home. Some Ptarmigan inhabit both continents, but there is no species of Grouse common to the Old and New Worlds.

North America appears to be the natural home of the Tetraonine, for, of the twenty-two known species, fourteen are found within her borders. Of these, three live on, and to the westward of the Rocky Mountains,—*Bonasa Sabinei, Canace Franklinii,* and *Lagopus Leucurus;* five dwell to the eastward of this great range—*Bonasa Umbellus* and *Umbelloides, Canace Canadensis, Cupidonia Cupido,* and *Pediocætes Phasianellus;* four are to be met with on both sides of the Mountains—*Centrocercus Urophasianus, Pediocætes Columbianus, Dendragapus Obscurus,* and *D. Richardsonii;* and two are inhabitants of the extreme northern parts of the continent—*Lagopus Albus* and *Lagopus Rupestris.* Europe possesses six species—*Bonasa Sylvestris, Lyrurus Tetrix, Tetrao Urogallus, Lagopus Scoticus, Lagopus Mutus,* and *Lagopus Albus.* Asia has four—*Bonasa Sylvestris, Tetrao Urogalloides, Falcipennis Hartlaubii,* and *Lagopus Albus. Bonasa Sylvestris* has also been found in Japan, and *Lagopus Hyperboreus* is peculiar to Spitzbergen.

Although not so brilliant or attractive in their plumage as the Pheasants, yet, in consequence of the delicacy of their flesh, the Grouse are valuable birds, and in the bleak regions of the frozen north, the Ptarmigan are one of the chief means of subsistence for the inhabitants, who kill thousands of them annually, and salt their flesh for the winter's consumption. Perhaps no family of birds, excepting the Phasianidæ, contains species of so much importance to man, as those comprising this Monograph, whether considered as affording him food, or as objects of sport in the field; and as many of them are capable of domestication to a certain extent, they may be introduced into the aviary, or among the inhabitants of the poultry-yard, where in many instances their gentle dispositions would make them desirable acquisitions.

Hybridism is of common occurrence among the members of this family. I have seen the offspring produced by the crossing of eight distinct species, and have no doubt but that wherever the territories inhabited by separate species join, the birds will mingle and breed together. These hybrids always bear characteristic markings by which it is comparatively easy to ascertain their parentage, and I have never heard of a single instance where the hybrids of two distinct species of Grouse have produced *inter se.* If such indeed was usual, the Tetraonine would soon consist of a confused mass of aberrant forms, from among which it would be impossible to extricate a single original species, and to prevent such an untoward result as this, nature has interposed by rendering hybrids, as a general rule, infertile. It is a well-known fact that hybrids between different species of Pheasants— *P. Colchicus, P. Torquatus,* and *P. Versicolor*—naturalized in England, have produced *inter se,* but to what extent I have no means at hand to enable me to state; but it is probable their confined boundaries, and semi-domesticated condition, may account in a great measure for their ability to breed in and in without the introduction of fresh blood. Instances have been recorded where hybrids between different species of the Anatidæ have produced *inter se,* but these may be deemed exceptional cases, for in one, at least, it was ascertained by dissection that this fertility did not extend to the second generation.[*] It is in these days, I think, established beyond controversy, that hybridism is of no unusual occurrence among gallinaceous birds in a wild state; but it seems necessary, in order that these hybrids should become fertile beyond the second generation, that they must at least be semi-domesticated; for the proof of which, we may look at the Pheasants as above cited. Why this should be so, is a problem of no easy solution; but probably the main cause is change of food, and to some degree of even their habits also, produced by their altered condition of life.

The Tetraonine may be divided into three groups—the Wood, Mountain, and Plain Grouse. For the first of these we have *Tetrao Urogallus, T. Urogalloides, Canace Canadensis, Canace Franklinii, Falcipennis Hartlaubii, Dendragapus Obscurus, D. Richardsonii, Bonasa Umbellus, Umbelloides, Sabinei, Sylvestris,* and *Lyrurus Tetrix.* In the next division are included *Lagopus Leucurus* and *Lagopus Mutus;* and for the third, or those species which habitually dwell upon the plains, we have *Centrocercus Urophasianus, Pediocætes Columbianus, P. Phasianellus, Cupidonia Cupido, Lagopus Albus, L. Scoticus, L. Rupestris,* and *L. Hemileucurus.*

At one period all the species of this family were included in the genus *Tetrao* established by Linnæus in 1735, having *T. Urogallus* as the type; but as they became better understood, it was found necessary to separate them into several genera, as the many distinct and varied forms presented themselves, so that now the genus *Tetrao* is restricted to the species *Urogallus* and *Urogalloides,* distinguished by the elongation of the feathers beneath the chin into a beard-like appendage.

In 1760 Brisson established the genus *Lagopus,* thus separating the Ptarmigan from the Grouse, and in 1819 Stephens included the species of Ruffed Grouse in the genus *Bonasa.* Mr. Swainson made a further division by instituting, in 1831, the genera *Centrocercus* and *Lyrurus* for the Cock-of-the-Plains and Black Grouse; and other changes have at various periods been made, until we now have no less than ten different genera, all of which seem to have furnished sufficient characters to warrant their having been established.

* A. Newton, Proc. Zool. Soc., 1860.

INTRODUCTION.

The Grouse are rather large in size, heavy in body, with small heads, the nasal fossæ filled with feathers concealing the nostrils; moderately long necks, short wings, rounded and concave beneath; stout legs and feet, the toes having pectinations of scales along the edges, the hind toe elevated above the plane of the rest; the tarsi covered with feathers, in the Bonasæ only halfway, in the Lagopidæ to the claws. I commence my review of the family with Mr. Stephens' genus.

BONASA,

WITH THE FOLLOWING CHARACTERS

Head crested, bill short, strong; wings short, concave beneath, third and fourth primaries longest; tail of eighteen broad feathers; lower half of tarsi naked, covered anteriorly with two rows of scales; sides of toes pectinated with scales; claws short and curved.

This genus has its representatives in both the Old and New Worlds, although the species inhabiting the former has not the ruff so developed as have those belonging to America. There is but little difference in the plumage of the sexes; the female being distinguished chiefly by the smaller size of the ruff; in the European species, by the absence of the black throat. The males are polygamous, and desert the females during the period of incubation. These birds go in flocks, and on being disturbed will frequently take refuge in trees. They are:—

This last species differs from the rest in being monogamous, although the male does not remain with the female while the latter is setting, and also in not possessing the peculiar habit of drumming, so characteristic of the other species.

The next contains only two members, the giants of the family, and to which Linnæus has given the term

TETRAO,

WITH THE FOLLOWING CHARACTERS.

Bill strong, upper mandible curved, head slightly crested, feathers of the chin elongated and pointed. Tarsi completely covered with hair-like feathers.

The forests of the Old World are the home of the magnificent species composing this genus; but in some localities where they were formerly abundant, they now exist in greatly reduced numbers; indeed, in some places, have become extinct. The only species are:

For the species composing the next group I propose the term

DENDRAGAPUS,

WITH THE FOLLOWING CHARACTERS.

Bill strong, upper mandible curved at tip; large air-sacs on each side of the neck, capable of inflation, but usually hidden by the feathers; wings rounded, third and fourth quills longest. Tail long, composed of twenty broad feathers. Tarsi feathered to the toes.

The birds included in this genus are inhabitants of the western portion of North America. They are of large size, the flesh white, and much esteemed as food. They are,

7.	DENDRAGAPUS OBSCURUS,	.	. PLATE VII.
8.	" RICHARDSONII,		. PLATE VIII.

For the fourth group, also consisting of only two species, I retain Reichenbach's genus

CANACE,

WITH THE FOLLOWING CHARACTERS.

Head without crest, neck destitute of air-sacs; tail long, of sixteen feathers.

These birds are natives of North America, dwelling in the thick parts of the forests; go in flocks; are generally of a tame, unsuspicious nature; their flesh dark and bitter. They are,

9. CANACE CANADENSIS,		. PLATE IX.
10. " FRANKLINII,		. PLATE X.

Mr. Boardman informs me that this species allows one to approach very closely to it when in the woods, without manifesting any alarm; and the only indication it gives of its intended flight, is by raising the membrane over the eye to its utmost extent, when the bird almost immediately takes wing, flying only, however, to a short distance.

For the next species, an inhabitant of Northern Asia, and remarkable for the peculiar formation of its primary quills, I propose the term

FALCIPENNIS,

WITH THE FOLLOWING CHARACTERS.

Head crested; wings short; the first four primaries greatly falcate; third and fourth longest. Tail moderate, of sixteen feathers. Tarsi thickly feathered.

The only one known is,

11. FALCIPENNIS HARTLAUBII,	. PLATE XI.

For the sixth genus, composed also of a single species, I retain Swainson's name,

LYRURUS,

WITH THE FOLLOWING CHARACTERS.

Bill strong. Wings moderate; the third quill longest. Tail very much forked; exterior feathers curved outwards. Tarsi feathered.

12. LYRURUS TETRIX, .   . PLATE XII.

For the next species I propose to employ Swainson's term of

CENTROCERCUS,

WITH THE FOLLOWING CHARACTERS.

Bill compressed. Base of culmen prolonged towards the crown of the head, dividing the frontal feathers. Tail long, of twenty feathers, which are lanceolate and pointed.

This bird is an inhabitant of the desert plains of western North America, and lives upon the Artemisia, which abounds in those regions.

13. CENTROCERCUS UROPHASIANUS, .   . PLATE XIII.

The next two species I include in the genus instituted by Prof. Baird,

PEDIŒCÆTES,

WITH THE FOLLOWING CHARACTERS.

Bill strong, moderate. Neck destitute of lengthened feathers. Wings short, rounded. Tail short, graduated, the upper middle coverts extending beyond the tail. Tarsi feathered to the base of the toes.

The Sharp-tail Grouse are dwellers of the plain, and are found in large flocks upon our western and northern prairies. They are,

14. PEDIŒCÆTES COLUMBIANUS,   . PLATE XIV.
15.   "   PHASIANELLUS,   . PLATE XV.

Hearne says of this last species: "These birds are always found in the southern part of Hudson's Bay, and are very plentiful in the interior parts of the country, and in some winters a few of them are shot at York Fort (lat. 57° north), but never reach so far north as Churchill. In color they are not very unlike the English hen-pheasant, but the tail is short and pointed, like that of the common duck; and there is no perceivable difference in plumage between the male and female."

"When full grown and in good condition, they frequently weigh two pounds: and though the flesh is dark, yet it is juicy, and always esteemed good eating, particularly when larded and roasted. In summer they feed on berries, and in winter on the tops of the dwarf birch and the buds of the poplar. In the fall they are tolerably tame, but in the severe cold more shy; frequently perch on the tops of the highest poplars, and will not suffer a near approach. They sometimes, when disturbed in this situation, dive into the snow; but the sportsman is equally balked in his expectations, as they force their way so fast under it, as to raise flight many yards distant from the place they entered, and very frequently in a different direction to that from which the sportsman expects. They, like the other species of Grouse, make their nests on the ground, and lay from ten to thirteen eggs. Like the Ruffed Grouse, they cannot be tamed, as many trials have been made at York Fort without success; for though they never made their escape, yet they always died, probably from the want of proper food, for the hens that hatched them were as fond of them as they could possibly have been had they been the produce of their own eggs. This species of Grouse is called by the Southern Indians, 'An-Kis-Cow.'"

For the next genus I retain Reichenbach's term,

CUPIDONIA,

WITH THE FOLLOWING CHARACTERS.

Bill moderate. Wings rounded. Tail short; the feathers stiffened. Gular sacs concealed by tufts of lanceolate feathers. Tarsi thickly feathered.

The single species of this genus, is peculiar to North America, and dwells in large flocks upon the western prairies. It is the

There seems to be some doubt among authors regarding the proper term for the remaining genus; some considering the Lagopus of Brisson as different from the Lagopus of Vieillot.

As the former is not deemed an authority for species, it may naturally be supposed that a like verdict would be rendered against him regarding genera; but in this instance he has instituted the genus, taking the Tetrao Lagopus of Linnæus as his type, and Vieillot, in his Analyse, has simply followed him.

It would therefore only be rendering due justice that I should retain for the next group Brisson's term of

LAGOPUS,

WITH THE FOLLOWING CHARACTERS.

Bill moderate; nasal groove covered with feathers, in winter reaching over on to the bill. Tail moderate. Legs and feet densely covered with hair-like feathers to the nails.

The species are,

INTRODUCTION.

From an examination lately made of this bird in the British Museum, I am obliged to state that in my opinion it is only a light-colored variety of Lagopus Scoticus, and therefore should be considered as merely a synonym of that species, and not in any way distinct.

In this recapitulation I have given, I believe, every known species among the Tetraoninæ, which may at the present day be entitled to a specific distinctness. As I intend always to keep the subject before me, I shall be happy at any time to receive any additional information regarding these birds, or to learn of the discovery of new species.

EXPLANATION OF THE ABBREVIATIONS

AND

LIST OF AUTHORS AND WORKS REFERRED TO.

Albin, Av. *Albin's Natural History of Birds.*

Aud., Ornith. Biog. *Audubon in the American Ornithological Biography.*

Aud., Syn. *Audubon in the Synopsis of Birds of America.*

Aud., Birds of Amer. *Audubon in Birds of America.*

Baird, U. S. P. R. R. Exp. Exped. *Baird in the Report of the Pacific Railroad Exploring Expedition.*

Baird, Birds of North Amer. *Baird in Birds of North America.*

Bart., Trav. in E. Flor. *Bartram's Travels in East Florida.*

Bechst., Naturg. Deutschl. *Bechstein, Gemeinnute Naturgesch Deutschlands.*

Bewick, Brit. Birds. *Bewick in History of British Birds.*

Ber., Hist. Prov. *Bernard's Mémoire pour Servir l'Histoire Naturelle de la Provence.*

Boie, Isis. *Boie in the Isis.*

Bon., Am. Ornith. *Bonaparte in the continuation of Wilson's American Ornithology.*

Bon., Mon. Tetrao., Trans. Am. Phil. Soc., N. S. *Bonaparte in a Monograph of the Tetraoninae in the Transactions of the American Philosophical Society. New Series.*

Bon., Geog. and Comp. List Birds. *Bonaparte in Geographical and Comparative List of Birds.*

Bon., Compt. Rend. *Bonaparte in the Compte Rendus.*

Bon., Rev. Ornith. Europ. *Bonaparte's Revue d'Ornithologie d'Europe.*

Briss., Ornith. *Brisson's Ornithologie.*

Brex, Birds of Eur. *Brex in Birds of Europe not found in the British Isles.*

Brehm, Vog. Deutschl. *Brehm in Vögel Deutschlands.*

Buff., Plan. Enlum. *Buffon's Planches Enluminées.*

Cooper and Suckl., Nat. Hist. Terr. *Cooper and Suckley in Natural History of Washington Territory.*

Cuv., Regn. Anim. *Cuvier's Règne Animal.*

Doug., Trans. Linn. Soc. *Douglas in Transactions of the Linnæan Society.*

Edw., Birds. *Edwards' Natural History of Birds.*

Elliot, Proceed. Acad. Nat. Scien. *Elliot in the Proceedings of the Academy of Natural Sciences of Philadelphia.*

Eyton, Cat. Brit. Birds. *Eyton in Catalogue of British Birds.*

Fab., Faun. Groen. *Fabricius' Fauna Groenlandica.*

Fab., Prod. des Isl. Orn. *Faber Prodromus der Island Ornithologie od. Geschichte der Vögel Islands.*

Gaim., Voy. en Scand. *Guimard's Voyage en Islande et Scandinavie.*

Gloger, Vog. Eur. *Gloger's Handbuch der Naturgesch der Vögel Europas, etc.*

Gmel., Syst. Nat. *Gmelin's Systema Natura.*

Gould, Birds of Eur. *Gould in Birds of Europe.*

Gould, Proc. Zool. Soc. *Gould in Proceedings of Zoological Society of London.*

Gould, Birds of Gt. Brit. *Gould in Birds of Great Britain.*

Grav., Brit. Ornith. *General British Ornithology.*

Gray, Gen. of Birds. *G. R. Gray's Genera of Birds.*

Gray, Cat. Birds Brit. Mus. *G. R. Gray in Catalogue of Birds in British Museum.*

Hahn, Vog. Deutschl. *Hahn's Vögel Deutschlands.*

Habel., Journ. für Ornith. *Hartlaub's Journal für Ornithologie.*

Hearne, Journal. *Hearne's Journal.*

Hearne, Voy. a l'Ocean du Nord. *Hearne dans une Voyage à l'Océan du Nord.*

Jard., Game Birds. *Jardine's History of the Game Birds.*

Jard. and Selb., Ill. Ornith. *Jardine and Selby's Illustrations of Ornithology.*

Jenyn., Man. Vert. Anim. *Jenyns' Manual of the British Vertebrate Animals.*

Keys. and Blas., Wirb. Eur. *Keyserling and Blasius' Wirbelthiere Europas.*

Kaup, Natur. Syst. *Kaup's Nature Systems.*

Lath., Ind. Ornith. *Latham's Index Ornithologicus.*

Leach, Syst. Cat. Mam. and Birds Brit. Mus. *Leach's Systematic Catalogue of the Mammals and Birds in the British Museum.*

Lewin, Brit. Birds. *Lewin's British Birds.*

Linn., Syst. Nat. *Linnæus' Systema Natura.*

Linn., Faun. Suec. *Linnæan Fauna Suecica.*

Meyer, Taschenb. Deutschl. *Meyer, Taschenbuch der deutschen Vogelkunde od. Kurze Beschreibung aller Vögel Deutschlands.*

McGill., Brit. Birds. *McGillivray's History of British Birds.*

Midden., Sibir. Reis. *Middendorf's Sibirische Reise.*

Mont., Ornith. Dict. and Supp. *Montgomery's Ornithological Dictionary and Supplement.*

Morris, Hist. Brit. Birds. *Morris' History of British Birds.*

Naum., Vög. Deutschl. *Naumann's Vögel Deutschlands.*

Newb., Zool. Cal. and Or., Route P. R. R. Surv. *Newbury on the Zoology of California and Oregon on the Route of the Pacific Railroad Survey.*

Nill., Faun. Suec. *Nilson's Fauna Suecica.*

Nutt., Man. Ornith. *Nuttall's Manual of Ornithology.*

Nill., Faun. Skand. *Nilson's Fauna Skandinavica.*

Ord, Guth. Geog., 2d Amer. Edit. *Ord in Guthrie's Geography, 2d American edition.*

Pall., Zoogr. *Pallas' Zoographica, Rosso-Asiatica.*

Penn., Arct. Zool. *Pennant's Arctic Zoology.*

Penn., Brit. Zool. *Pennant's British Zoology.*

Ray, Syn. *Ray's Synopsis Avium.*

Reich., Av. Syst. Nat. *Reichenbach, Systema Natura.*

Rich., Appen. Parry 2d Voy. *Richardson in the Appendix to Parry's Second Voyage.*

Ross, Arct. Exp. *Ross' Arctic Expedition.*

Say, Long, Exp. Rocky Mts. *Say in Long's Expedition to the Rocky Mountains.*

Selby, Brit. Ornith. *Selby's British Ornithology.*

Schleg., Rev. Crit. des Ois. d'Eur. *Schlegel, Revue Critique des Oiseaux d'Europe.*

Spar., Mus. Carl. *Sparrmann, Museum Carlsonianum.*

Steph., Gen. Zool. *Stephens' General Zoology.*

Suckl., Proc. Acad. Nat. Scien. *Suckley in Proceedings of the Academy of Natural Sciences of Philadelphia.*

Swain. & Rich., Faun. Bor. Amer. *Swainson & Richardson's Fauna Boreali Americana.*

Temm., Pig. et Gall. *Temminck, Histoire Naturelle générale des Pigeons et des Gallinacés.*

Thomp., Nat. Hist. of Irel. *Thompson's Natural History of Ireland.*

Vieill., Nouv. Dict. d'Hist. Nat. *Vieillot, Nouveau Dictionnaire d'Histoire Naturelle.*

Will., Ornith. *Willoughby. The Ornithology.*

Wils., Am. Ornith. *Wilson's American Ornithology.*

Yarr., Brit. Birds. *Yarrell's British Birds.*

LIST OF PLATES.

RUFFED GROUSE
BONASA UMBELLUS.

BONASA UMBELLUS.

RUFFED GROUSE.

TETRAO UMBELLUS. Linn., Syst. Nat., vol. i, p. 275 (1766).—Gmel., vol. i., p. 782.—Wils. Am. Ornith., vol. vi. (1812), p. 46, pl. 49.—Bon. Obs. Wils., 1825, p. 182.—Aud. Ornith. Biog., vol. i., 1831, pp. 211 and 260, pl. 41.—Id. Syn., 1839, p. 202.—Id. B. of Am., vol. v., 1842, p. 72, pl. 293.

TETRAO (BONASIA) UMBELLUS. Bonp., Syn. 1828, pp. 126.—Id. Mon. Tetrao, Am. Phil. Trans., vol. iii., 1830, p. 389.—Nutt., Man., vol. i., 1832, p. 657.

TETRAO TOGATUS. Linn., vol. i., 1766, p. 275.—Forst., Phil. Trans., lxii., 1772, p. 393.

TETRAO TYMPANUS. Bart., Trav. in E. Flori., 1791, p. 290.

BONASA UMBELLUS. Steph. Shaw's Gen. Zool., vol. xi., 1824, p. 300.—Bonp., List., 1838.—Id., Compt. Rend., xlv., p. 428.—Baird, B. of N. Am.—G. R. Gray, Cat. B. B. Mus., Pt. III, p. 46, 1844.

BONASIA UMBELLUS. Bonp., Geog. and Comp. List. B., p. 43, No. 282.—Id., Syn. (1828), p. 126.

This fine species, known in different localities by the respective names of Partridge and Pheasant, is one of the handsomest in appearance of the Grouse family. Graceful in its movements, it walks with a firm, proud step, erecting its head, and opening its tail with a quick, sudden jerk.

The Ruffed Grouse is widely distributed, as it is found from Maryland northward throughout the eastern part of the United States, and westward to the Rocky Mountains. It becomes scarcer in Virginia, and does not exist in South Carolina, at least in the maritime districts. The males are polygamous, and abandon the females when incubation commences, associating in small parties by themselves until the autumn, when they join the hens, and old and young birds remain together until spring. The flight of this species is straight, and very rapid, but not usually protracted to any great distance. It rises from the ground with a prodigious whirring of the wings, and after proceeding by quick flapping until under full headway, continues its course by sailing, and generally alights in some thick clump of bushes.

The most peculiar habit of the Ruffed Grouse is that of drumming, and it is usually practised in the spring, although the strange sound produced by this custom may be heard in the summer and fall, sometimes as late as November. Early in April, the male resorts to some chosen log, every morning soon after dawn, and again towards sunset, and is accustomed to strut up and down with head drawn back, tail expanded to its fullest extent, and wings lowered and buzzing against the bark. After a few moments passed in this way, he suddenly stops, and stretches out his neck, draws the feathers close to the body, lowers his tail, and beats his sides violently with his wings, increasing the rapidity of the stroke at every movement. The sound produced by this action is not unlike the rolling of distant thunder, and may be heard a considerable way off. As soon as the females hear this noise they fly directly to the spot, and it is not uncommon for several hens to be gathered around the male at one time, admiring his gallant bearing as he thus parades before them. The male, unless disturbed, will resort to the same log throughout the season; and these places are easily recognizable by the quantity of feathers and excrement lying around.

The nest, composed of leaves and plants, is placed upon the ground, and contains from ten to twelve yellowish eggs, sometimes spotted with dull red; and these frequently become the spoil of some hungry crow, as the female rarely covers them when she leaves her nest. The mother evinces the greatest affection towards her young, which follow her as soon as they are hatched, and she tries by every means in her power, feigning lameness, etc., to draw away the attention of her enemies from the helpless brood in order to cause pursuit to be made after herself. In this she is generally successful; and when she has drawn her pursuer to what she may consider a safe distance from her young, she suddenly takes wing, and returns by a circuitous flight to the spot from which she was disturbed.

The Ruffed Grouse feeds upon seeds and berries of all kinds, and also upon the leaves of several species of evergreen. Late in the winter, if the snow has been deep, or of long continuance, they eat the leaves of the Kalmia Latifolia, and their flesh becomes very bitter and disagreeable; sometimes it is even dangerous to be eaten. They roost in trees, generally choosing the places where the foliage is thickest, taking up their positions at a little distance from each other. When suddenly startled by a dog or other animal, they will often take refuge in the nearest tree, and stand upright close to the trunk, where they will remain so motionless that it requires a practised eye to discover them. The flesh of the Ruffed Grouse is white, delicate, and highly esteemed as an article of food; and when half grown, these birds are eagerly sought after, for unfortunately there is no dish more in demand in August than chicken Partridges; and although in some States the fine is very heavy for killing them at this season, yet great numbers are destroyed.

The usual resort of this species is the craggy hill-side, and the rocky borders of streams, where the foliage is dense, and the bushes very closely grown together. In the autumn they will leave the mountains, and go down into the warmer temperature of the swamps to pass the winter. These birds have many enemies: various species of hawks are always ready to pounce upon them; while foxes, coons, weasels, etc., destroy both them and their eggs.

Mr. George A. Boardman states, that this species is in the habit in winter of sleeping under the snow, and frequently, on account of a crust forming during the night, through which they are unable to penetrate, very many are imprisoned and perish from starvation.

The Ruffed Grouse may be described as follows:

Head and back part of neck, yellowish-red; back, deep chestnut, interspersed with white spots margined with black; tail, reddish-yellow, barred and mottled with black, with a broad subterminal band of the latter color; a bar through the eye, yellowish-white; throat and lower part of breast, brownish-yellow. The feathers of the ruff, which are always most conspicuous in the male, are velvet-black, with blue reflections; under parts white, with large spots of brownish-red; under-tail, coverts mottled with the same; bill, horn-color, black at tip; lower part of tarsi and feet, brown. A great difference is observable in specimens, some being of a grayish hue, and with gray tails. This variation, I think, does not prove that there are two species, but merely varieties of the typical form; as it is often the case that the eggs in the same nest will produce both styles of coloring.

The plate represents a male upon his log, with a hen surrounded by her brood, observing his proud attitude. The figures are all life-size.

ARCTIC RUFFED GROUSE,
BONASA UMBELLOIDES.

BONASA UMBELLOIDES. ELLIOT.

ARCTIC RUFFED GROUSE.

TETRAO UMBELLOIDES. Doug., Linn. Trans., vol. xvi., 1829, p. 148.
BONASA UMBELLOIDES. Elliot, Proceed. Acad. Nat. Scien. (1864), p.

This Grouse, so closely resembling the common species generally known as Ruffed Grouse that a casual observer would be likely to confound them, seems, however, to be entitled to specific distinctness. In the specimens before me differences exist, which are constant, and of such a character as leave me no alternative but to separate this bird from *Bonasa Umbellus*. In the first place, the size is very much less, at least one third, and there appear none of the red hues, so very conspicuous in some specimens of the Ruffed Grouse. Also (and this is a constant character in the specimens I have), the broad black band crossing the lower part of the tail is not continuous, but is broken by the two centre feathers, which retain their gray color spotted with black, throughout their length. The markings of the back, in shape and distribution, differ materially from its ally, and the ruff is not nearly as conspicuous as is that in *B. Umbellus*.

Douglas, in his Paper on the Grouse in the Transactions of the Linnæan Society, speaks of this bird, under the head of *T. Umbellus*, as follows:—"In the valleys of the Rocky Mountains, 54° north latitude, and a few miles northward near the sources of Peace River, a supposed variety of this species is found,—different from *T. Umbellus* of Wilson. On comparing my specimens from that country with some which I prepared in the States of New York and Pennsylvania, and on the shores on the chain of lakes in Upper Canada, I find the following differences: First, the northern bird is constantly one third smaller, of a very light speckled mixed gray, having little of that rusty color so conspicuous in the southern bird; secondly, the ruffle consists invariably of only twenty feathers, these short, black, and with but little azure glossiness; the crest feathers are few and short. Should these characters hereafter be considered of sufficient importance for constituting a distinct species, it might perhaps be well to call it *T. Umbelloides*." It seems to separate the eastern from the western species, for while the former possesses both the gray and red varieties, the latter has only the red, and the present species but the gray. It is distributed from the South Pass of the Rocky Mountains northward throughout the entire range, and on the slopes as far as the woods extend. It has also been found eastward to the shores of Slave Lake.

The species may be described as follows: Upper part of head and neck brownish gray, with central feathers of the crest black, crossed with irregular bars of rufous brown, a white line from the bill running to and under the eye, with spot of same behind and rather above the eyes. Feathers above the ruff, of a darker shade than the head, broadly marked with black, a central strip of white, sometimes widening at the tip. Ruff moderate, glossy black, with purple reflections. Upper part of back barred with black, and rufous, these crossing but not including the shaft, which is reddish brown. Rest of back and upper tail coverts light gray mottled with black, each feather having a black spot terminating in a yellowish-white, irregularly heart-shaped spot. These are indistinct upon some of the feathers, and the black spot only shows through the gray color at intervals. The upper wing coverts are reddish, with central streaks of white, these last predominating, giving a very light appearance to this portion of the bird. Wings darker than the back, each feather with a central line of white, and the tertials spotted with black, this last being quite conspicuous on the inner webs of some, while the outer webs have very broad lines of white next the shaft, and separated from the brownish gray of the outer portion by a narrow line of dark brown. Spurious wings dark brown, shafts reddish brown. Primaries same color, but the outer webs have alternate marks of yellowish white and brown. Tail light gray, irregularly crossed by narrow, interrupted bars of black, and mottled also throughout the entire feather with the same; a broad band of black crosses the tail near the tip, but is interrupted by the two central feathers, which preserve their gray hue throughout their length. Throat white, spotted with brown on the sides, a narrow band of rufous, spotted with black, crosses the upper part of the breast. Rest of under parts white, the feathers crossed with bars of dark brown, most distinct on the flanks. Under tail coverts dark gray irregularly marked with faint lines of black, and having very broad white ends. Under part of tail feathers of a silvery gray, less distinctly mottled and crossed with black than the upper side. Upper mandible black; under mandible horn color at base, tip black. Tarsi brownish white. Feet brown.

The plate represents the two sexes of the natural size.

RUFFED GROUSE.

BONASA SABINEI. BAIRD.

SABINE'S GROUSE.

TETRAO SABINEI. Doug., Trans. Linn. Soclet., vol. xvi. (1829), p. 130.—Swain. & Rich., Faun. Bor. Amer., vol. ii., p. 343 (1831).
TETRAO UMBELLUS. Newb., Zool. Cal. & Oreg. Route. Rep. P. R. R. Surv., vol. vi., p. 94 (1857).
BONASA SABINEI. Baird, U. S. P. R. R. Exp. & Surv., vol. ix. p. 631.—Ib. Birds of North America, p. 630.—Coop. & Suckl., Nat. Hist.
    Wash. Territ., p. 224.—Elliot, Proc. Acad. Nat. Scien., 1864.

THIS handsome bird, an inhabitant of the western portion of our continent, is very common on the coast of Washington and Oregon Territories, and also in Vancouver's Island.

It resembles in its habits the Ruffed Grouse of the more eastern States, and frequents wooded and mountainous districts; but on account of the dense cover in which it chiefly remains, it is approached with difficulty.

In the spring, the drumming of the male may be heard in the early morning, summoning the hens into his presence. This noise, resembling the rolling of a distant drum, is produced by rapid and violent beating of the wings, and can be distinguished a considerable way off. The females soon assemble, and no doubt greatly admire the pompous bearing of their lord, as he struts with expanded tail before them.

The species is polygamous; the male deserting the females during the period of incubation, and leaving the young brood entirely to their watchful care. The nest is placed in some thicket, the better to conceal its contents from the prying eyes of some thieving crow or raven, either of which have a decided weakness for making a meal upon the eggs.

The young run as soon as they are hatched, and, at the slightest note of alarm from their vigilant mother, squat, and lie so still and close to the ground as to render it no easy undertaking to discover them.

Douglas was the first to constitute this bird as distinct from the common Bonasa Umbellus, on account principally of its dark red color, and the absence of any of the gray hues so prevalent in the eastern species. Prof. Baird, in the Birds of North America, is also inclined to consider it entitled to specific distinctness, basing his opinion not only on the color of the plumage, but also upon the great length of the middle toe.

If this last character was confined to the western bird exclusively, it would undoubtedly, in conjunction with the difference in color, be a good reason for giving it a specific value; but as an equal length of the toe can also be found in the Ruffed Grouse, it would seem best not to take that into consideration; and therefore this bird's claims for separation would rest upon the color of its plumage.

At the present time, it would seem that no specimen has been obtained, among the varying examples of the Bonasa Umbellus, on the eastern coast, which presents the deep rich hues of the typical form of Bonasa Sabinei; and as the Bonasa Umbelloides, an apparently good and distinct species, inhabits an intermediate region, it is perhaps best to retain the western form under the appellation given to it by Douglas, rather than to consider it merely a variety of the common B. Umbellus.

Still, if at some future period examples should be procured west of the Rocky Mountains, possessing the different variations, from these deep red colors, to the light gray so perceptible in some specimens of the Ruffed Grouse, the conclusion would be a natural one to consider the birds inhabiting both sides of the continent as but one species; although the singular fact would remain, that they were divided by a different and smaller species.

I regret that sufficient material from the west coast has not yet been obtained to settle this question satisfactorily.

The flesh of Sabine's Grouse is white, tender, and well flavored, in no way inferior, I believe, to that of its eastern relative; and as the forests of those distant regions are gradually thinned by the axe of the hardy pioneer, and the pursuit of these birds is rendered less difficult, then undoubtedly they will become as much an object of attention to the sportsman as are the Ruffed Grouse in our more thickly populated States at the present day.

My plate represents the two sexes, of the natural size; and they may be described as follows:

General color dark orange chestnut, mottled upon the back and wings with black, each feather having a distinct central streak of reddish white. Head and neck lighter than the body; flanks reddish yellow, barred with black, and having the central marks of reddish white. Primaries dark reddish brown, mottled on the outer webs with reddish yellow. The tail, dark red, is tipped with gray, with a subterminal bar of black, beyond which is another line of gray, followed by eight or ten irregular narrow bars of black. The under tail coverts are orange chestnut indistinctly barred with black, terminating with an angular white spot. Tufts on sides of the neck dark metallic green. Feathers on the thighs and tarsi reddish gray. Bill dark brown, feet yellowish.

HAZEL GROUSE,
BONASA SYLVESTRIS.

BONASA SYLVESTRIS. Steph.

HAZEL GROUSE.

TETRAO BONASIA. Linn., Syst. Nat., vol. i., p. 275 (9th edit ?).—Gmel., Syst. Nat., vol. i., p. 753.—Lath., Ind. Ornith., vol. ii., p. 640.—
    Bree, Birds of Eur., vol. iii., p. 203.
TETRAO CANUS. Nills, Fauna Skania.
LA GELINOTTE. Buff., Plan. Enlum., pl. 474, 475.
LA GELINOTTE. Puula des Coudriers, Cuv., Règ. Anim., vol. i., p. 448.
TETRAS GELINOTTE. Lieu., Pig. et Gall., vol. iii., p. 174.
TETRAO BETULINUS. Scop. Ann., i., No. 172.—Gmel., Syst. Nat., vol. i., p. 749.—Lath., Ind. Ornith., vol. ii., p. 637.
BIRCH GROUSE. Lath., Syn., iv., p. 735, 5.
HAZEL HEN. Will., Birds (Ang.), p. 176.
BONASIA EUROPÆA. Gould, B. of Eur., pl. 251.
BONASIA SYLVESTRIS. Bon., Rev. Ornith. Eur., p. 174.
BONASA SYLVESTRIS. Gray, Gen. of Birds, vol. iii.—Ib., Cat. Birds, Brit. Mus., Part. III., p. 46 (1844).—Bon., Geog. and Comp. List
    Birds, p. 43, No. 292.—Steph. Shaw, Gen. Zool., xi. (1819).—Elliot, Proc. Acad. Nat. Scien. (1864).

THE Hazel Grouse is the only representative of this genus found in the Old World. It is pretty generally distributed throughout Europe and Asia, its range extending from France on the West, through Sweden, Norway, and Germany, as far as the River Lena in Siberia, and, according to Dr. Schrenck in his recent "Reisen und Forschungen in Amur Land," it is common at all seasons of the year from the southern coast of the Okhotsk Sea to the Bay of Hadschi, and also in the island of Saghalien, as well as from the mouth of the Amoor to its source in Dauria. In Great Britain it is not found. It frequents the birch and pine forests, and, like its American relatives, is partial to the sides of hills and mountains.

The species is monogamous, and does not, I believe, possess the peculiarity of drumming, like our birds of this genus; and in these two particulars lie the principal differences in their habits. They rise from the ground with the same loud whirring noise, but their flight is not generally continued to any great distance.

The breeding season commences in April, and the sexes separate as soon as incubation commences; the males keeping by themselves leaving the brood to the care of the female, but returning when the young are about half grown. The usual number of eggs is ten, of a yellowish color spotted with brown. Their flesh, like all of this genus, is white and tender, and is generally considered among the most delicate of the Grouse family.

The Hazel Grouse is rather a small bird, approaching nearest in size to our B. Umbelloides, and is destitute of the ruff so conspicuous in all the other species. In Dree's "Birds of Europe," in the article upon this species, is a translation from Dr. Schrenck's work above mentioned, a portion of which I take the liberty of inserting here:

"* * * * Scarcely any locality can be named where it is not found, yet it appears principally in the north of the Amoor, on the borders of rivers in the mixed forests of birch, aspen, poplar, alder, and willow bushes, and in the south principally in the light-foliaged woods and the underwood which grows along the rocky banks of the rivers. Not unfrequently, also, I have met with it, in winter and summer, on the willow-grown islands, or on such shores as those of the Amoor, Gorin, and Ussuri. In as great numbers did I find the Hazel Grouse in the wildest parts of the Amoor Land, where it was by no means shy. In the Nikolajev Posten, and on the River Tyrny, in Saghalien, I have been able to shoot several times at a pair of individuals in a tree before the others flew away. In Saghalien, and on the Gorin, they flew up before us, and kept in a circuit round about us. In summer, when the noise of our movements roused them, they often settled down on a tree close by the river, enabling us to shoot them from our hiding places. They were among the daily contents of our game bag in the Amoor Land, where, as well as in the Bay of Hadschi and the snow fields of Saghalien, they gave us as good sport as in the light and sunny oak hedges on the Ussuri. In the summer of 1855 I found a nest with eggs on the borders of the Lake of Kidsi. It was in a fir wood, at the foot of a tree, concealed in the moss and brushwood. The eggs were of the usual dark yellow, with many brown spots and points, and were hatched on the 14th of June. On the 26th of July I met with a family just fledged at Puchale, near the mouth of the Gorin, in the leafy underwood of a pine forest. The moulting of the Hazel Grouse takes place at Nikolajev Posten in August and September. On the 23d of August I found the moulting far advanced, and every wing and tail feather freshly grown. It was quite concluded on the 1st of October."

BONASA SYLVESTRIS.

Captain Blakiston mentions in the "Ibis" for 1862, p. 329, that he "brought from Japan a single young male grouse, which Dr. Sclater considered to be of this species, and which Mr. Maximovitch, who had killed them, pronounced to be identical with those of the Amoor. This is" (he goes on to say), "I believe, the first instance of this bird being found in Japan; probably it does not inhabit the more southern part of the empire. As far as I saw, it has the same habits as the Ruffed Grouse of North America." On another occasion four fine males with black throats were killed, but these he was unable to save.

The male may be described as follows: Top of head, neck and shoulders rufous brown, barred with black; back and rump lighter brown mottled with black, and each feather edged with gray. Chin and throat black, the latter edged with a broad band of white. A white mark also before and behind the eye. A band of rufous across the fore part of breast, each feather with a white streak near the tip. A broad white band in front of wings. Wing coverts reddish brown, mottled with black, some feathers having central streaks of white, widening at the tip, others with round white spots. Under parts white, irregularly marked with black or brown. Secondaries and primaries brown, the former tipped with reddish yellow, the latter having their outer edges mottled with reddish yellow and brown. Tail gray, confusedly mottled with black, and having a broad black band near the tip. Tarsi half covered with grayish hairy feathers, the naked parts brown. Bill black. Feet brown.

The female differs from the male in having the throat yellowish brown, and a reddish brown spot before the eye.

The figures are of life size.

TETRAO UROGALLUS.

COCK-OF-THE-WOODS.

TETRAO UROGALLUS. Linn., Faun. Succ., No. 200.—Id., Syst. Nat., vol. i., p. 273.—Gould, B. of Eur., pl. 248.—G. R. Gray, Gen. of
B., vol. iii.—Graves, Br. Ornith., vol. ii.—Lewin, B. Birds, vol. v., pl. 133.—MacGilliv., B. Birds, vol. i., p. 136.—Gmel., Syst.
Nat., vol. i., p. 746.—Temm., Man. d'Ornith. (1815), vol. i., p. 265.—Brehm, Vög. Deuts., p. 501.—Naum., Vög. Deuts. (1833),
vol. vi., p. 277, t. 154 and 155.—Jenyns, Man. B. Vert. Anim., p. 164.—Keys and Blas, Wirbleth. Eur., p. 64.—Schleg., Rev.
Crit. des Ois. d'Eur., p. 75.—Gray, Cat. B. B. Mus., Pt. III., p. 45 (1844).—Bon., Geog. and Comp. List B., p. 43, No. 293.—
Id., Rev. Ornith. Eur.—Elliot, Proc. Acad. Nat. Scien. (1864).
COQ DE BRUYÈRE. Buff., Plan. Enlum., pl. 73 and 74.
LE GRAND COQ DE BRUYÈRE. Cuv., Reg. Anim., vol. i., p. 448.
UROGALLUS MAJOR. Briss., vol. i., p. 182.—Buff., Hist. Prov., vol. ii., p. 331.—Germ. Ornith., vol. ii., p. 83, t. 236, 237.
MUSCOVIAN BLACK GAME-COCK. Albin., vol. ii., pl. 29 and 30.
WOOD-GROUSE. Morris, Hist. of B. Birds, vol. iii., p. 328, pl. 103.—Brit. Zool., vol. i., No. 92, t. 40, 41.—Penn, Arct. Zool., vol. ii., p.
312.—A. Supp., p. 62.'—Tour in Scotl., 1769, p. 217.—Lath., Syn., vol. iv., p. 729.—Id., Ind. Ornith., vol. ii., p. 634.—Mont.,
Ornith. Dict. and Suppl.—Penn, Brit. Zool., 1812, vol. i., p. 348. pl. 44, 45.—Thomp., Nat. Hist. of Irel., p. 31.
TETRAO CRASSIROSTRIS. Brehm, Vog. Deuts., p. 504.
TETRAO MACULATIS. Brehm, Vög. Deuts., p. 504.
TETRAO MAJOR. Brehm, Vög. Deuts., p. 503.
THE CAPERCAILLIE. Yarr., Brit. B., 2d edit., vol. ii., p. 325, fig.

The following synonymy is that which has been given to a hybrid between this species and the Lyrurus Tetrix:

TETRAO HYBRIDUS. Linn., Faun. Succ., p. 72.—Sparr., Mus. Carl., t. 15.—Gould, B. of Eur., pl. 249.—Gray, Geo. of B., vol. iii.—
Lath., Ind. Ornith., vol. iii., p. 636.—Temm., Man. d'Ornith., 1815, p. 287.—Naum., Vög. Deuts. (1833), vol. vi., p. 304, t.
156.—Penn, Brit. Zool. (1812), vol. i., p. 355.—Flem., Brit. Anim., p. 41.—Gray, Cat. B. B. Mus., p. 141 (1850).
TETRAO MEDIUS. Jenyns, Man. B. Vert. Anim., p. 162.—Brehm, Vog. Deuts., p. 506.—Yarr., B. Birds, 2d edit., vol. ii., p. .—Bon.,
Comp. and Geog. List B., p. 43, No. 294.
TETRAO INTERMEDIUS. Langed., Mem. l'Acad. Peters., vol. iii., t. 14.—Keys and Blas., Wirbleth. Eur., p. 64.

This magnificent species—the largest of all the members of this family—is a native of the Old World; and, on account of its size and
splendid appearance, has been well called the King of the Game-Birds. It is found in considerable numbers in Prussia, Austria, Switzer-
land, Norway, Sweden, and Russia as far north as Siberia; and at one time was quite plentiful in Scotland, where, however, at the present day, it has become very rare—indeed, in most parts extinct. Repeated efforts have been made, by the owners of large estates, to
reintroduce it, by importing the birds from Norway; but it is very doubtful if the Cock-of-the-Woods will ever again become abundant
in its former island-home.

It remains always in the vicinity of the pine and fir trees, upon the leaves of which it feeds, and loves to stay in the depths of the
lonely forests, where, amid the dense undergrowth, it is concealed from every eye.

The flight of the Capercailzie is rather heavy, and the rapid beating of its wings produces a sound which may be heard for a considerable distance. When upon the ground—where, during the summer, it passes much of its time—it carries its tail drooping, and its head
well forward, presenting rather a dull appearance.

This species, like many others of this family, is polygamous, and the male deserts the females when incubation commences; the young
remaining with the hen generally throughout the winter.

In the spring the cock is accustomed to utter his call-note from the branch of some tree where he has passed the night. His man-
uvers at this time are very eccentric and peculiar; and the following account, taken from Boner's "Forest Creatures," gives a very vivid
description of the way in which the male is accustomed to summon the hens into his presence:

TETRAO UROGALLUS.

"Toward morning—but long before dawn—at two or three o'clock generally, and while the stillness yet reigns upon the earth like a superincumbent thing, the Cock awakes and begins his peculiar call. Though low in tone, such an absolute quiet reigns, that it is heard distinctly even when you are not close to the bird.

"Before this begins, however, you must be near the tree you noted so well the preceding evening. As it is still night, there is some difficulty in discovering any object; and only the dark, undefined outlines of large masses like trees can be discovered as you peer upward, and your vision grows accustomed to the darkness.

"But hark! from a distance you hear a sound which, did you not know what it was, you surely would never interpret. From a tree-top it comes across to you through the air, sounding something like a person pronouncing 'Tut, tut,' gutturally, in the *depth* of his throat, or as if two pieces of hard wood were being knocked against each other.

"Well, that's a cheering circumstance; for, though you knew he must be there, you were not sure if he would call or not, and without that there were no possibility of approaching him. And after rising at midnight, and a walk of some hours through the wood, and a cold hour's watching before the dawn, it is vexatious to hear nothing; and still more so, when day is just breaking, to distinguish the dark form of the capercaile a hundred yards distant, on a projecting bough.

"But this morning there is no cause for regrets, or lamentation, or complaint. You are at your post betimes, the bird is not far, and he has begun his love-call; and that is all you can desire. He repeats it often, too, and quicker, and more quickly, and you have a foreboding of success; for such accelerated utterance betokens that the sweet frenzy possesses him, and that love and its madness are blinding him, even as they blind men. The guttural 'tut, tut,' is followed by another, not unlike the smack *with the tongue* one curious in wine will give after having tasted a sort which he finds superlatively excellent. This is repeated a few times, and then comes a changing, now louder now lower, sound, resembling a long drawn-out 'whish,' or that gliding sound which a scythe makes in sweeping at morning through the heavy, dewy grass. This is the close of the call; and while he utters it, he spreads out his tail like a fan, the wings, quivering with excitement, are extended downward, and with head outstretched, and all the feathers round the neck standing on end like a ruff, he pirouettes on his perch, or goes sideways to and fro the whole length of the branch. It is during this finale that the bird may be approached; for while the fit is on him, while the ecstasy lasts, he sees and hears nothing."

When uttering this note, the hunter takes a few steps forward, and then remains motionless, waiting for a repetition of the call; and if he times his movements properly, he is enabled to approach within shot, without having his presence noticed by the bird.

Early in May, the female, having selected a place amid long grasses, or in the thick bushes, for her nest, which is carelessly formed, lays from eight to twelve eggs, and after about four weeks' hatching the young appear. These immediately desert the nest and follow the hen, who evinces the most tender care for them, and feeds them upon ants' eggs, insects, &c.

This species is the type of the genus Tetrao, as constituted by Linnæus, and which formerly was made to include nearly all the members of this family; but, as now restricted, it contains only this bird and its relative, the *T. Urogalloides*.

A great difference exists in the relative size of the two sexes of the *T. Urogallus*, the male being nearly as large again as the female; and her flesh is much more preferable for food, as it is tender and juicy.

The male has the entire upper parts blackish brown, every feather speckled with grayish. Head and neck similarly marked; the feathers of the throat which are elongated are black. Breast black, with rich green reflections on the upper portion. Flanks brownish gray speckled with black, and with a few white feathers intermingled. Upper tail coverts like the back, tipped with white; under coverts black, also margined with the same. Tail black. Bill horn-color; feathers on the legs brown, with a few bars of dark brown. Feet brown.

The female has the upper parts a rich reddish brown, barred and blotched with black; the feathers of the hind neck and rump tipped with grayish white. Sides of neck, and throat and breast, rich orange, barred with black on the former; rest of lower parts lighter orange, each feather tipped with white. Tail reddish brown, barred with blackish brown. Bill dull horn-color. Tarsi covered with grayish brown feathers. Toes brown.

The average length of the male is about three feet, and its weight eight or nine pounds. The female is about two feet in length.

The plate represents the two sexes, considerably reduced in size.

ALLIED COCK OF THE WOODS

TETRAO UROGALLOIDES. Middend.

SIBERIAN WOOD GROUSE.

TETRAO UROGALLOIDES, Middend., Siber. Reis., Band. II.—Elliot, Proc. Acad. Nat. Scien. (1864).
TETRAO UROGALLUS—var. MINOR. Pallas, Zoogr. R. A., II., pp. 58, 59.

This Wood Grouse is a native of Northern Russia, Siberia, and Kamtschatka, and is fully described by Middendorf in the work above referred to. He says: "Pallas notices that Messrs. Schmidt had distinguished a smaller variety of Wood Grouse; Steller, on his part, I find, affirms that the Wood Grouse of Kamtchatka were obviously smaller than those of Siberia and Russia. These statements are explained by the fact that in Siberia I have met with two varieties of the Cock-of-the-Wood, i. e., the one considered as typical in Europe, and a smaller variety or species living in the mountainous regions; and this latter appearing to be the only one existing in Kamtschatka. * * * * Our species has several distinguishing characteristics, which we will review in order. First, in regard to the male. It is considerably smaller than the European Cock-of-the-Woods, the full-grown bird weighing probably not more than 9½ pounds at the highest estimate. This small size is very apparent on a comparison of the bills and the toes. The following measurements of a large male of the typical form, and of the Tetrao Urogalloides of the Stanowój Mountains show the difference existing:

	Tetras Urogallus.	Tetrao Urogalloides.
Height of bill, . .	32 millimetres.	21 millimetres.
Breadth of upper mandible, .	24 "	17 "
Length of upper mandible, .	30 "	26 "
" " gap, ...	50 "	44 "
" " middle toe without nail, ...	65 "	62 "

"By this the great difference in the bill of the two forms will be seen. Notwithstanding this, the difference in the size of the whole body is not very great, for on account of the proportionately longer tail of Tetrao Urogalloides, its entire length is equal to that of Tetrao Urogallus. With this last species, the tail when laid on the back generally reaches to the beginning of the neck; with the other, however, it goes to the back of the head. This is in consequence of the length of the middle feathers, which give to the tail a wedge shape.

"The upper parts of Tetrao Urogalloides are always covered with white spots; and those upon the upper coverts of the wings and tail are characteristic of this species." * * * *

There is a fine example of a male of this species in the British Museum, with even more white spots upon the wings than were upon those of the specimen figured in my plate; and in the Paris Museum a very well preserved female. This last resembles closely the female of Tetrao Urogallus, but the wings and back were very much spotted and marked with white.

This species resembles in some degree its near relative, the Tetrao Urogallus, but may easily be distinguished from it by the rows of white spots upon the wings, and also sometimes on the back, its small size, and wedge-shaped tail.

The figures are about three-fourths the natural size.

DENDRAGAPUS OBSCURUS. Elliot.

DUSKY GROUSE.

TETRAO OBSCURUS. Say. Long. Exp. Rocky Mts., vol. ii., 1833, p. 14.—Bon. Syn., 1828, p. 127.—Ib. Mon. Tetrao, Am. Phil. Trans., vol. iii., 1830, p. 391.—Ib. Amer. Ornith., vol. iii., 1828, Pl. xviii.—Newb. Rep. P. R. R. Surv., vol. vi., 1857, p. 93.—Gray, Gen. of Birds, vol. iii.—Baird, P. R. R. Exp. and Surv. Report. Zool., vol. ix., p. 620.—G. R. Gray, Cat. Birds, Brit. Mus., Part iii., p. 46 (1844).—Bon. Geog. and Comp. List Birds, p. 43, No. 283.—Coop. and Suckl. Nat. Hist. Wash. Territ., p. 218.

CANACE OBSCURUS. Bonp. Compt. Rend., vol. xlv., 1857, p. 428.

DENDRAGAPUS OBSCURUS. Elliot, Proceed. Acad. Nat. Scien. (1864), p.

The Dusky, Blue, or Pine Grouse, by either or all of which names it is known, is next in size to the Cock-of-the-plains, of the American portion of this family, and like that species is also an inhabitant of the Western part of the United States. But it differs from the Sage Cock, which is a prairie-loving bird, in making its home amid the mountains and dense spruce forests. It is very abundant in the main chain of the Rocky Mountains, the Black Hills of Nebraska, the Cascade Mountains of Oregon, and thence to the Pacific, wherever the country is wooded sufficiently to afford it shelter.

The male, like several other species of American Grouse, possesses the power of inflating a neck on each side of its neck, and producing a mournful sound by exhausting the air in the same. Thus, in the spring, where these birds are plenty, this peculiar call may be heard on every side, and, like the drumming of the Ruffed Grouse (Bonasa Umbellus), it seems to possess the power of ventriloquism; for should you seek the bird, guided by the noise, you would probably discover that it came from quite a different direction from that apparently indicated by the sound.

In November the Dusky Grouse are generally missed from their accustomed haunts, and will not be met with again, save perhaps now and then a straggler, until the following spring. This disappearance has given rise to many theories among the inhabitants of the regions in which it dwells; one of which is similar to that formerly entertained of the swallow, that they pass the winter in a state of torpidity, not, however, in this case, in the mud, but among the thick-clustering foliage of the evergreens. It is a well-known fact, that the Ruffed Grouse, as the winter grows severe, leaves the mountain sides, where it has perhaps passed the summer, and descends to the warmer temperature of the thick swamps, there to remain until the ice melts under the rays of the returning sun. And without doubt the present species also leaves its summer haunts, and either descends to the milder climate of the valleys, or migrates to a limited extent southward.

My friend Dr. Geo. Suckley, in his Natural History of Washington Territory, gives the following interesting account of the disappearance and habits of this Grouse:

* * * * "In the autumn, about November 15th, they generally disappear, and it is rare indeed to see a single individual of the species during the interval between that period and about March 20th of the following year. Concerning the whereabouts of this bird during the winter, there are many opinions among the settlers. Some maintain that the species is migratory, and that they retire to the south, while others say that they repair to the tops of the highest evergreen trees, where, in the thickest foliage of the branches, they pass the cold season in a state of semi-torpor, rarely or never descending until warm weather comes on. As they subsist well on the leaves of the coniferæ, and can always obtain sufficient water from the snow and rain drops on the leaves to supply their necessities, I have but little doubt that this latter is the correct account, or that, if migratory, they are but partially so.

"I saw one bird of this species on the ground during a fall of snow, in January, 1854, near the Nisqually River, Washington Territory; and I have been told that a man near Olympia, Washington Territory, whose eyesight is excellent, is able any day during the winter to obtain several birds by searching carefully for them in the tops of the tallest and most thickly leaved firs. This requires eyesight of greater power than most men possess.

"Even in the summer, when these birds are generally lower in the trees, it is very difficult to find them among the dense branches. They have, in addition to their sombre hues, the advantage of their habit of crowding very closely to the limbs, and of sitting almost immovably for hours. The first indication, in the spring, of their arrival (?) or activity (?), is the courting call of the male. This call is a prolonged noise, sounding much like the whir of a rattan cane whirled suddenly through the air. It is repeated quickly several times, and then stops for a brief interval. This noise is said to be produced by inflating and contracting a couple of sacks on each side of the throat, which are for the most part concealed when collapsed, and are covered by an orange-yellow, thick, corrugated, unfeathered skin.

DENDRAGAPUS OBSCURUS.

"These birds, at Fort Steilacoom, are very abundant throughout the spring and early summer. They are there mostly confined to the forests of fir trees (*Abies Douglasii*). Late in the season, after hatching, they may be found generally at midday on the ground, in search of berries, seeds, &c. When alarmed, they almost invariably seek safety among the dense foliage of the trees, instinctively appearing to understand the advantages of thus hiding. In the autumn they are more generally found on the ground, feeding on sallal and other berries. One day in October, 1856, I saw on the Nisqually plains, among fern and grass, five of these birds, full grown, and in excellent order. A man killed the whole five, one by one, with a double-barrelled gun, without an attempt being made by a single individual to fly. This Grouse is a very fine table bird; the little dash of *pine taste* its flesh possesses only adding to its game flavor. I have known males, in June, weighing three and a half pounds, although they rarely exceed two and three fourths pounds. By August 1st, the young are generally half grown. They are then easily killed on the wing, and are excellent for the table."

The flesh of this Grouse is white, resembling in appearance that of the Ruffed Grouse (*B. Umbellus*), to which, in the consideration of many, few birds can compare, as regards tenderness and flavor. When on the ground, it will lie very close, sometimes starting up almost from under your feet, and generally, instead of seeking safety in distant flight, will take refuge in the nearest tree, where it will remain as motionless as the branches themselves, and in this manner escape, since it is next to impossible to discover it, as stated in the passage I have quoted above. The male exceeds the female in size, and is almost unequalled in beauty of plumage and gallant bearing, among the American Grouse. Its geographical distribution appears to be Northern California, on the coast as far as the Columbia River, as far as the coast of Oregon and Washington Territories, and thence southward in the main chain of the Rocky Mountains as far as Texas.

As this species, together with its near relative, commonly known as *Tetrao Richardsoni*, appear to possess sufficient characters to distinguish them from the genus *Canace* (a term formed to include the American Wood Grouse), in having gular sacks, and tail composed of twenty feathers, I have deemed it best to include them in a separate genus by themselves, and have therefore proposed the term *Dendragapus*, or *Tree-loving*. The nest is formed upon the ground, and the eggs are of an ash-brown color. The male has the entire upper parts of a leaden gray, each feather mottled with rufous brown and black, this color extending throughout the upper tail coverts, the two middle feathers of which are tipped with ashy.

The wings are bluish gray, mottled similarly to the back, with, however, larger spots and bars, and inclined to ashy near the end of the feathers; the primaries and greater portion of the secondaries brown, with their outer webs of a light brown. Space before the eye, chin, and throat, white, irregularly crossed with black. Breast and abdomen dark lead color; the feathers on the flanks broadly marked with white (in some instances, with a white central streak widening at the end). A spot of white upon the neck just forward of the wing. This covers the naked skin of the gular sacks, when it is not inflated. The tail feathers are black, rounded at the end, with a broad terminal band of ash gray, and the under coverts dark lead color, broadly tipped with white. Thighs and tarsi pale brown. Bill black.

The female has the upper parts of a grayish brown, each feather with bars of black and rufous brown; the black bars broadest and most conspicuous upon the lower part of the neck and back; and here also are bars of brownish yellow in place of the rufous brown. Upper part of head yellowish brown, crossed with five dark brown lines; back of neck leaden gray, indistinctly barred with black lines; upper tail coverts grayish, with zig-zag lines of black and yellowish brown. Wings lighter brown than the back, but similarly crossed with brown and black, and the shafts of the feathers whitish. Primaries and secondaries black, the outer webs of both mottled with a very light brown, darker, however, on the secondaries. Tail black, excepting the central feathers, which are marked like the back, and with a broad terminal band of ash gray. Throat white, faintly marked with brown; upper part of head dark lead color, with irregular lines of yellowish brown crossing near the end of the feathers; under parts lead color, lighter than the male, and much obscured with white, and the feathers bordering the belly broadly tipped with white. Under tail coverts dark gray, crossed with black lines, and tipped with white. Thighs and tarsi light brown. Bill black.

The plate represents the male of life size, and a reduced figure of the female in the distance.

RICHARDSON'S GROUSE,
TETRAO RICHARDSONI.

DENDRAGAPUS RICHARDSONII. Elliot.

RICHARDSON'S GROUSE.

TETRAO OBSCURUS. Aud., Ornith. Biog., vol. iv., 1838, p. 446, pl. 365.—Id., Syn., 1839, p. 283.—Id., B. of Amer., vol. i., 1842, p. 89, pl. 295.—Natt., Ornith., vol. i., 1840, p. 609.
DENDRAGAPUS RICHARDSONII. Elliot, Proc. Acad. N. S. (1864).
TETRAO RICHARDSONII. Doug., Linn. Trans., xvi., p. 141.

Resembling very closely the Dusky Grouse, the present species would probably be considered by an ordinary observer as identical with that bird; yet it presents characters which vary from its ally, and which are constant, and of sufficient value to constitute a specific distinctness.

Richardson's Grouse is strictly a mountain species, never, to my knowledge, having been observed on the plains, and in its habits presents no material difference from its relative. It is an inhabitant of the Rocky Mountains, and is found from the South Pass northward.

The most striking difference between this species and the *Dendragapus Obscurus*, is in having the tail square at the tip, of a uniform black throughout its length, and being entirely destitute of the ashy terminal band so conspicuous in its ally.

Audubon figures this species, in the "Birds of America," under the name of *Tetrao Obscurus*, but speaks of one specimen in his possession which had the tail considerably rounded, of a deep black, with a terminal band of ash-gray. He did not consider these as different, but accounted for the variation by supposing "that when the tail is unworn it is distinctly rounded, and tipped with gray."

This does not appear to me to be the case, as the tails of the specimens before me present no indication of being worn away at the tip, and the feathers are broader than those of its relative.

The upper parts of the male are grayish brown mottled with light brown; in some specimens this mottling is wanting, the feathers being of a uniform color; upper tail coverts tipped with gray. Wings light brownish gray mottled with brown, the secondaries margined with whitish; primaries light brown. Head, neck, breast, and abdomen lead color. Chin and throat white, irregularly crossed with black. Patch of feathers before the wing white. Flanks bluish gray, many of the feathers tipped with white. Tail black, square at the tip, and inclining to brownish on the outer webs. Under coverts dark brown broadly margined with white. Thighs and tarsi pale brown faintly mottled with a darker brown. Bill black. Feet brown.

The female has the upper parts lighter than the male, covered with bars and blotches of blackish brown. Head and neck grayish, similarly crossed with brown. Wings rufous brown mottled with blackish, some of the feathers having a central white streak. Primaries light brown, the outer webs mottled with yellowish brown; secondaries margined with grayish white. Throat and breast crossed with dark brown. Under parts lighter than the male, with considerable white intermingled. Tail black, the central feathers mottled like the back. Under coverts dark brown margined with white. Bill black. Feathers of the legs light brown.

Over the eyes of both sexes is an orange-colored membrane.

The plate represents a male and female, the latter reduced in size.

SPRUCE GROUSE.

CANACE CANADENSIS. REICH.

SPRUCE GROUSE.

TETRAO CANADENSIS. Linn., Syst. Nat., vol. i., p. 274 (1766).—Forst., Phil. Trans, lxii. p. 389 (1772).—Gmel., vol. i., p. 749 (1788).—Sab., Zool., Appen., Frank. Exp., p. 683.—Bon., Syn., p. 127 (1828).—Ib., Am. Ornith., vol. iii., pl. xxi. fig. 2 (1829).—Ib., Am. Phil. Trans., vol. iii., N. S., p. 391 (1830).—Sw. & Rich., Faun. Bor. Amer., vol. ii., p. 346 (1831) pl. lxii.—Aud., Ornith. Biog., vol. ii., p. 437 (1834).—Ib., Syn., p. 203.—Ib., Birds of Amer., vol. v., p. 83, pl. 294 (1842).—G. R. Gray, Gen. of Birds, vol. iii.—Baird, Birds of Amer., p. 622 (1860).
CANACE CANADENSIS. Reich., Av. Syst. Nat., 1851, Type.—Bon., Comptes Rendus, xlv., p. 428 (1857).—Elliot, Proceed. Acad. Nat. Scien. (1864).
TETRAO CANACE. Linn., Syst. Nat., vol. i., p. 275 (1766).
BLACK SPOTTED HEATHCOCK. Edw., pl. cxviii.
SPOTTED GROUSE. Penn.
LA GELINOTE NOIRE D'AMERIQUE. Cuv. Règ. Anim., vol. i., p. 449.

SCATTERED throughout the northern United States to the Arctic Sea, and westward nearly to the Rocky Mountains, this Grouse is found amid the solitudes of the spruce forests or swamps, making its home amid their deepest recesses, where man rarely intrudes, or where, from the depth of the treacherous moss, which covers the swamp with a mantle of green, he is unable to pass. The Black Partridge, by which name this bird is sometimes known, is generally tame and unsuspicious, and, unlike the majority of the members of this family, does not seem to stand much in dread of man's presence. They are easily tamed, and appear to bear confinement well, readily feeding upon oats, wheat, and other kinds of grain. They commence to breed in the United States about the middle of May, and farther north nearly a month later. The female conceals her nest, which is composed of leaves and moss, under the drooping branches of the fir tree, and lays from ten to fourteen eggs, of a deep buff color, spotted with brown.

The males leave the females at the commencement of incubation, and betake themselves to a different part of the forest, and remain there until late in autumn, when they join the females and young. During the period that they are thus alone, they are more shy and wary than at any other season of the year. In the spring the males strut before their mates with the tail expanded to its utmost extent, and wings lowered to the ground; at intervals springing into the air and beating their sides. As an article of food, the flesh of this Grouse is dark and disagreeable, being frequently so bitter as to render it unfit to eat; although I believe they are more palatable when they feed solely on berries.

The chicks represented in the plate were obtained for me in Maine by Mr. Geo. A. Boardman, a gentleman much devoted to the science of ornithology. They were taken by the Indians employed by my friend, who were compelled to exercise considerable patience in their capture, for these young are so nimble and rapid in their movements, and hide so expertly at the first warning note from the female, that it is no easy matter to catch them; the difficulty of pursuit being greatly increased by the density of the forests they inhabit, and the slippery, miry nature of the ground, into which a foot, if not very cautious, would frequently sink to his waist.

The usual appearance of the male is as represented in the plate, but I have specimens in my cabinet which have almost the entire breast black. This may possibly be the result of age. When this species is started, it generally flies but a short distance, and takes refuge in some thick spruce tree, where it will remain motionless, watching its pursuer, and is easily shot upon its perch.

The male may be described as follows:

Upper parts plumbeous gray, each feather crossed with bars of black parallel to each other; wings and flanks reddish brown, mottled similarly to the back; secondaries tipped with yellowish white; primaries dark brown, the outer edges mottled with yellowish brown; upper tail coverts lighter than the back, mottled with black, and tipped with gray. Throat and pectoral band black; the former edged with a white band—a white mark also before the eye; under parts white, crossed irregularly with black; tail dark brown, with a terminal band of orange chestnut; under tail coverts black, barred and tipped with white; bill black; legs covered with hairy feathers of a yellowish brown; feet brown. ,

The female is much lighter than the male, upper parts similarly barred with black, but mixed with orange; flanks, sides of the neck, and wings brownish orange, crossed with black,—the feathers of the wings having a central streak of white widening at the tip; primaries and secondaries brown, marked on the outer edge with yellowish brown; throat yellowish white; centre of abdomen white, barred with black; tail dark brown, crossed with five or six rows of reddish orange, and broadly tipped with the same; feathers of the thighs and tarsi yellowish brown; bill blackish brown; feet brown. Over the eyes of both sexes, a conspicuous vermilion membrane.

The young are of a lemon yellow, darker on the breast; a black bar through the eye; top of head and wings rufous brown, irregularly marked with black; upper mandible black; lower, light horn color; feet pale flesh color.

My plate contains the first representations of these that have been given.

CANACE FRANKLINII. Elliot.

FRANKLIN'S GROUSE.

TETRAO FRANKLINII. Doug., Trans. Linn. Soc., vol. xvi., 1829, p. 139.—Swain and Rich., Faun. Bor. Amer., vol. ii., 1831, p. 348, pl. lxi.—Baird, U. S. Ex. Exp. P. R. R., vol. ix., p. 623.—Coop. and Suckl., Nat. Hist. Wash. Territ., p. 221.—Nutt., Man. Ornith., vol. i., 1832, p. 667.
TETRAO CANADENSIS. Var. Bon. Am. Ornith., vol. iii., 1830, p. 47, pl. xxi. ♂.—Ib. Syn. 1828, p. 127.
TETRAO FUSCA? Ord. Guth. Geog., 2d Am. edit., vol. ii., 1815, p. 317.
CANACE FRANKLINII. Elliot, Proceed. Acad. Nat. Scien. (1864), p.

Until very recently, considerable doubt has been entertained by ornithologists, whether or not the present bird was a variety of the common Spruce Grouse (*C. Canadensis*). Prince Charles Bonaparte, in his continuation of Wilson's Ornithology, gives a figure of the male of this species, which came from the Rocky Mountains, and makes some comparisons between it and our well-known bird, closing his remarks by disclaiming that he should be understood as insinuating that there were two different species. With the limited materials that were at that eminent ornithologist's command at the time he wrote his article, it was very natural that he should hesitate to separate these birds, since the difference in their plumage might possibly have been (to use his own words) "entirely owing to season, though it is asserted that this species does not vary in its plumage with the season."

Within a short period, however, the Smithsonian Institution has, through its collectors, come into possession of specimens of both sexes of this Grouse, and the differences in their plumage are as characteristic and constant as are those by which the Spruce Grouse is verified. Professor Baird, in his article on this species, contained in the ninth volume of the Pacific R. R. Report, was satisfied of the specific distinctness of these birds; although he had only mutilated skins upon which to form his judgment, yet "the difference from *Canadensis*, however, even in these, is sufficiently appreciable." The species do not differ much in size, but if there is any, Franklin's Grouse is a little the larger of the two, but the structure of the tail feathers is quite different; those of *Canadensis* being much narrower and rounder at the end, while those of the present bird retain their width the entire length, being square, and, if anything, rather wider at the tip. The female also differs in the color of her plumage from that of *Canadensis*, being of a richer brown on the breast, and in having the tail and upper tail coverts tipped with white.

Dr. Suckley, who obtained specimens of this Grouse, says that it is "abundant in the Rocky and Bitter Root Mountains, also found in the Cascade Mountains, Washington Territory, near the Yakima Passes. This bird by the Indians has the jargon name 'Tyee Kulla Kulla,' or the 'chief bird,' or perhaps more correctly, the 'gentleman bird.' The specimens of *Tetrao Franklinii* sent by me to the Smithsonian, were obtained by Lieut. J. Mullan, U. S. A., at the St. Mary's Valley, in the Rocky Mountains. Lieut. Mullan stated to me that they were quite an abundant bird in that region, and very readily killed, as they are tame and unsuspicious."

Mr. Douglass, in the Linnean Transactions, gives the following short account of this species. He says: "Its flight is similar to the last mentioned ('Ruffed Grouse'); the present, however, runs over the shattered rocks and among the brushwood with amazing speed, and only uses its wings as the last effort of escape. Next on the ground, composed of dead leaves and grass, not unfrequently at the foot of decayed stumps, or by the side of fallen timber in the mountain woods. Eggs 5 to 7, dingy white, somewhat smaller than those of *Columba palumbus*. I have never heard the voice of this bird, except its alarm note, which is two or three hollow sounds, ending in a yearning, disagreeable, grating noise, like the latter part of the call of the well-known *Numida Meleagris*. It is one of the most common birds in the valleys of the Rocky Mountains, from latitude 50° to 54°, near the sources of the Columbia river. It may perhaps be inhabit higher latitudes. Sparingly seen in small troops on the high mountains which form the base or platform of the snowy peaks, 'Mount Hood,' 'Mount St. Helen's,' and 'Mount Baker,' situated on the western parts of the continent. In habit the present species assimilates more with *T. Canadensis* than any other. The unusually long square tail, constantly tipped with white, as is also the case with the upper and under coverts of the tail, are characters too prominent to be overlooked."

I would add here, in reference to Mr. Douglass' statement that the "tail is constantly tipped with white," that I have never seen that character in any specimen which has come under my observation. The tail feathers invariably retained their uniform black to the end, and it was only upon the upper and under coverts that the white was visible.

CANACE FRANKLINII.

The male has the upper parts black, each feather crossed with waving lines of london gray; the scapulars and wings are of a reddish brown, mottled irregularly like the back; the primaries dark brown, with reddish brown shafts; the upper tail coverts black, mottled with rufous, and broadly tipped with white; tail uniform black, the feathers broader than that of *Canadensis*, and nearly square at the tip. The throat, upper part of breast, and centre of belly pure black; a white line from the eye continues around the black of the throat. Feathers on the sides of the breast and belly, with conspicuous bars of white. The flanks, barred and mottled like the wings, the feathers having a white line in the centre expanding toward the tip. Under tail coverts black, broadly tipped with white. Thighs and tarsi light brown, faintly barred with darker brown. Bill black.

Female. Entire upper parts gray, barred with black and orange yellow, those on the head and neck being narrow, but becoming broader and conspicuous on the upper part of the back; upper tail coverts gray crossed with black, and but faintly with orange yellow, the outer feathers tipped with white, and running more than half way down the tail. Wings reddish brown, irregularly barred with black; some of the secondaries have a central line of white, widening at the top. Primaries brown, with the outer webs yellowish brown. Throat white, spotted with black. Upper parts crossed with black and orange yellow, the feathers tipped with white, this last becoming more prominent on the belly and flanks. Tail black, conspicuously barred with dark orange yellow, and tipped with white. Under coverts black crossed with yellow, and broadly tipped with white; so that when they are in position one above the other, it appears as though the coverts were white. Thighs and tarsi ashy brown, with marks of white appearing through the feathers on the former. Bill black.

HARTLAUB'S SPRUCE GROUSE.
FALCIPENNIS HARTLAUBI.

FALCIPENNIS HARTLAUBII. ELLIOT.

SIBERIAN SPRUCE GROUSE.

TETRAO CANADENSIS, var FRANKLINII. Midden., Siber. Reis., band. ii., theil. 2.
TETRAO FALCIPENNIS. Hartl., Journ. für Ornith., vol. iii. (1855), p. 39.
FALCIPENNIS HARTLAUBII. Elliot, Proc. Acad. Nat. Scien. (1864).

This species was discovered by Middendorf, who described it in the work referred to above, as Tetrao Canadensis, believing it to be the same as our Spruce Grouse; but Dr. Hartlaub obtaining some specimens, at once perceived them to be different, and named it Tetrao Falcipennis.

Middendorf says: "I first saw this bird on the Ujan, that is, right among the steep spurs of the Stanowoj Mountains. It is of very frequent occurrence on these slopes, and particularly in the neighborhood of Udskoj-Astrog."

"From this region, the first intelligence of this bird was received as narrated by Steller, according to whom, it was called by the inhabitants of Yakoutsk the wood-cock of the mountains."

* * * * "Our bird occurs on the entire southern coast of the sea of Ochotsk, and also all over the Stanowaj Mountains, and even over the southern slopes of the same as far as the region of the Shilka. Thus, this wood-hen is found on the Tiski, on the sources of the Kilé, and even on the middle portion of the course of the Ur; although it is not seen on the Linna or on the Oidé, which may be considered as forming its most southern limit."

"How far north this species may be found is unknown to me, but Woznesenky met this bird near Ajan, and I conversed with inhabitants of Yakutsk who had seen it on the road between Yakutsk and Ochotsk."

This Grouse bears some resemblance to the Canace Franklinii, but has many characters to distinguish it from that species, as a glance at the plate will testify; but it differs from the members of the genus Canace, by having the first four primaries, falcate; this being the sole instance among the Gallinaceous birds where this peculiarity is found, excepting the species of the genus Penelope.

As this is such a very marked and unusual occurrence, I have deemed it worthy in this instance of generic distinctness, and have therefore proposed the term Falcipennis; and in compliment to the eminent ornithologist who first detected this bird as of a distinct species, have given to it the name Hartlaubii, which I sincerely trust it may always be permitted to bear.

The very spirited drawing of Falcipennis Hartlaubii, which adorns this work, is the production of Mr. Wolf's pencil, and gives a perfect representation of the bird in its native wilds.

The female, as described by Middendorf, bears some resemblance to that of Canadensis, but "has the yellow of the throat and upper part of the breast more extended and spotted like the back, and is without the brown spots on the end of the tail."

THE BLACK GROUSE.
LYRURUS TETRIX.

LYRURUS TETRIX. Swain.

BLACK GROUSE.

TETRAO TETRIX. Nill., Faun. Suec., No. 202.—Linn., Syst. Nat., vol. i., p. 272, sp. 2.—Gould, Birds of Eur., pl. 250.—Jard. & Selb., Ill.
    Ornith., pl. 53, 47.—Pall., Zoogr., vol. ii., t. 52.—Gray, Gen. of Birds, vol. iii.—Gmel., Syst. Nat., vol. i., p. 748.—Lath., Ind.
    Ornith., vol. ii., p. 633.—Graves, Brit. Ornith., vol. ii.—McGill, Brit. Birds, vol. i., p. 145.—Leach, Syst. Catal. Mam. & Birds,
    Brit. Mus., p. 27.—Flem., Brit. Anim., p. 43.—Naum., Vog. Deuts., vol. vi., p. 324, t. 157 (1839).—Jenyn's Man. Brit. Vert.
    Anim., p. 166.—Eyton, Catal. Brit. Birds, p. 30.—Keys & Blas., Wirb. Eur., p. 64.—Bon., Geog. & Comp. List Birds, p. 44,
    No. 295.—Brehm, Vog. Deuts., p. 510.—Selby, Brit. Ornith., pl. 58., p. 423.
COQ DE BRUYÈRE À QUEUE FOURCHUE. Buff., Plan. Enlum., pl. 172, 173.—Ib., Hist. Prov., vol. ii., p. 536.
UROGALLUS MINOR. Raii Syn., p. 53, A, 2.—Will., p. 124, t. 31.—Briss., vol. i., p. 186.—Albin, vol. ii., t. 34.
TETRAS BERKHAN. Temm., Pig. et Gall., vol. iii., p. 140.—Ib., Man. d'Ornith., vol. i., p. 461 and 289 (1815).
GABEL SCHWANZERGES. Wahlbaim, Beobst. Naturg. Deut., vol. iii., p. 1319.—Meyer, Taschenb. Deut., vol. i., p. 295.
TETRAO DERBIANUS. Gould, Proc. Zool. Soc., p. 132, 1837.—Gray, Gen. of Birds, vol. iii.
BLACK GROUSE. Penn., Arct. Zool., vol. ii., p. 314.—Will., p. 173, t. 31.—Lewin's Brit. Birds, t. 134.—Montg. Ornith. Dict. & Supp.—
    Pult. Cat. Dorset, p. 7.—Don., Brit. Birds, 4, t. 97.—Bewick's Brit. Birds, vol. i., p. 298 (1797).—Penn., Brit. Zool., vol. i.,
    p. 352, pl. 46.—Morris, Hist. Brit. Birds, vol. iii., p. 335, pl. 170.—Thomp., Nat. Hist. of Ireland, vol. iii., p. 34.
UROGALLI'S TETRIX. Kaup., Natur. Syst. p. 180.
TETRAO JUNIPERORUM. Brehm, Vog. Deuts., p. 309.
LYRURUS TETRIX. Swain., Faun. Bor. Amer., p. 497.—Bon., Rev. Ornith. Europ., p. 174.—Gray, Catal. Birds, Brit. Mus., p. 142
    (1850).—Elliot, Proc. Acad. Nat. Scien. (1864).

This fine bird is distributed generally throughout the northern portions of Europe and Asia, but decreases in numbers as you go toward the south. Although it dwells in the large forests, in places where the birch tree grows, and among the juniper bushes, yet it prefers the moors and plains. Its food consists of ants' eggs, beetles, insects, and various kinds of berries, and in winter the young shoots of plants, to get at which it scratches away the snow from beneath the trees.

The male is a noble-looking bird, walks with a considerable strut, and has a very independent air; while the fine steel blue of his plumage, with the scarlet rings around his eyes, give to him a very attractive appearance. The wings are quite short, but its flight is rapid, and often well sustained. The Black Grouse is fond of the society of its own species, and generally they live together in small flocks or families: the old males, however, prefer to remain aloof from the rest, excepting in the spring. They are wild and quick-sighted, and it is difficult to approach them unobserved.

The pairing season commences about April, and each male usually has a chosen piece of ground to which he resorts every morning to associate with the hens, and also to engage in battle with some rival. At this time they are exceedingly pugnacious, and their conflicts are fierce, and prolonged until the weaker is driven away. Sometimes, not satisfied with gaining victories on his own territory over all invaders, the black cock will make excursions into the domains of his neighbors, to seek new conflicts with them.

To see them, at this time, it is necessary to be astir before the day has begun to dawn. At this early hour, the fluttering of wings and the peculiar chuckle in the woods notify us that the black cock is about to seek his mate, and he soon alights in some open ground. The hen gives notice of her presence by a low, uncertain note, uttered from her perch in some tree close by. The habits of the male at this particular time are very curious and eccentric; for sometimes five or six will meet together in the same trysting place, and, while night holds her sway and the sun has not yet gilded the snowy peaks of the loftiest mountains, they go dancing around seemingly a charmed circle, and flutter about as though held by some mystic spell. As each new comer arrives, he utters a low cluck, and joins in the curious antics. These round dances are interrupted every moment by several of the birds engaging in a desperate struggle, during which they spring into the air and beat their wings rapidly, uttering quick and angry clucks. In a work lately published by Mr. Charles Boner, entitled "Forest Creatures," is a very interesting account of this species during this period, as witnessed by the author. I give the article in his own words:

LYRURUS TETRIX.

"But in order to be exact, the following details are given of an excursion to Bohemia for the purpose of shooting Black Cock, as well as the experience then gained of this animal's peculiarities:

"As we had far to go, we left our inn betimes, and, the forester preceding us with a lantern, on we went behind each other through the coppice and the low grounds, where formerly there was a lake, but which lately had been drained. At this season the fields and moor land were all under water, and for an hour and a half we went splashing through the inundated plain. At night and in the fog it was difficult not to miss the usual landmarks, and to avoid the trenches cut to carry off the floods. After groping about at the spot where the huts made of fir boughs were erected, we saw them at last looming through the vapor, and each of us took his station in that assigned him. At this place, be it observed, the ground was not under water, though shaky and very marshy. To be out at early morning and to listen to the gradual awakening of animal life around, and to hear how the very earth seems to be shaking off its deep slumber, and at last to see forms appearing in masses, and, gradually taking well-known shapes, emerge from the gloom—this is one of the most interesting incidents among the very many which form the sum of a hunter's life.

"For a short time after arriving in the hut all was still as death. First was heard the low, sad cry of the goat-sucker—earliest of birds—as he flew through the darkness over the marsh; and presently, from the skirts of the wood, came the bleat of the roe that had been startled by a sound, or not improbably had caught the taint of our presence as a breath of air began to stir the leafless brambles on the dry spots around. The cry of a scared animal thus heard amid the profound stillness is very startling. It makes the same impression as of a man talking in his sleep. Presently the faint chirping of the water lark was audible; of the coot, and other dwellers in the morass. But now came a cheery sound, foretelling that the sun was about to appear, and that he—that rejoicing singer—was going forth to meet and watch him come. Straight over head rose a lark, pouring forth his gladdening song; and, accustomed as we are to hear the bird when we can look up and follow him on his heavenward flight, it did seem strange to listen to his warbling now, while no light as yet was in the air. Then from a distant village came the lugubrious 'Toot! toot!' of the watchman's horn, and a clock announced it was past three. Again the sharp bleat of a roe, but this time from a meadow in the direction of the hamlet. There is now on all sides an awakening; there is a hum in the water, and in the air, and in the woods, at first low and indistinct and tremulous, but gradually growing in volume, and becoming stable and definite. Now a snipe calls, and now a covey of partridges in fluttering flight whir by. There is a sound of waters everywhere oozing, yet rather felt than heard, it is so low and stealthy—not separate, but mixing with, and part of, the murmur of nature around. The blackness is changed into a confused gray; but hark! there is a fluttering and a rush of wings, which tells most surely that a cock has come to the trysting place. And now another rushing of pinions, and the same low 'Cluck! cluck!' as before. You look through the branches of your hut in the direction whence the sound proceeds, and peer into the gloaming. But it is not yet possible to distinguish anything. However, you hear the rush and the flutter of new comers; you hear, too, the half-cooing, half-clucking tones they utter, rising and falling by turns, as they give expression to their passionate longing. Then follows a sudden and rapid beating of wings, and quick and sharp angry cluckings; for the joust has already begun, and they are fighting wrathfully. How you long to see what is going on, and to behold the manœuvres which you well know that fluttering betokens! And now they are clucking quite near, and there is a violent beating of wings as they bound upward in their strife a few feet from the ground. If the haze would but disperse that you might get a shot! When suddenly from one of the huts, where your comrade is stationed, comes the report of a gun, which tells you that yonder is less mist than here, or that the birds being nearer enabled him to fire. But now you too are able to see something, and about one hundred and fifty yards off there is a black cock in the grass. To the right is another, and now from behind a hillock a third emerges. What can they be about? With outstretched neck they move creepingly onward, with a sort of would-be gravity, and then stand still in the same position as before, looking as ridiculous as possible. But presently they begin dancing up in the air, and turning round like a turkey cock, the tail feathers erect and outspread. Up they jump again a foot or two, clucking and gobbling the while; and then they will suddenly resume their old posture, and, poking out their neck to its fullest stretch, move mincingly forward, and with affected gait. But they approach each other now, and a fight ensues, and the weaker is driven away. They are still pretty far, but a rifle bullet may hit one still. Your sights are fine—necessarily fine—and it is not day yet; however, you try, and the sharp crack of the explosion rings through the neighboring wood. By Jove! there is the very fellow at which you aimed exactly where he was; he is looking up, it is true, somewhat surprised, but a moment more and he is at his old tricks again, creeping along as silly as before. It reminds you of the 'medicine man' in Catlin's 'Indians,' who is playing just such antics as our black cock here, whom we have come a day's journey to see. He calls in a somewhat coaxing tone, and the three notes of which his invitation consists are indicative of impatience and longing. Another shot from your comrade's gun, but it does not disturb them. They go on dancing in a ring as before. It is a laughable sight. And now turning on the opposite side of your hut, you look what is to be seen there, and behold! another 'medicine man' is having his dance. Does the distance, as viewed through your peep-hole, deceive you, and is he not within range of your gun? It was too far, for the bird runs a dozen yards as if a shot or two had touched him, and then stalks and jumps and pirouettes as before. And yonder are three, four, five, six more, but far off and beyond reach of mine or my comrade's gun. Now they come hopping along like boys jumping in sacks; and they may at last be within range; but now they stop and go off in another direction, with their necks made as long as possible, poking close to the ground. One flies to the lower branches of a young birch, and chuckles inwardly at the recollection of his wooing. Presently he takes wing, and you watch him making for the forest; but you tell yourself he will be there again to-morrow, and there is satisfaction in that certainty. One after the other flies away, for it is day now, and you are glad to emerge from your shelter and move your benumbed limbs; and though there is a two-hours' walk before getting home, and half of it is wading through water, still there is a warm breakfast in perspective, and that is at all times cheering.

"From the other hut comes my comrade; and what has he shot? There lie six fine cocks as the result of his morning's work. And how did he manage it? With the exception of one bird, all came close to where he was, and they made his task an easy one. To-morrow they might fall more in the other direction, and then that would equalize our sport.

"It is always a chance whether the birds come in the immediate neighborhood of your retreat, or close enough for a shot. But what does not happen one morning may the next. And this watching and expectancy have their charm. Nor while you are waiting

2

and hoping are you without amusement. The time does not seem long while observing their habits and drollery. On the snow such dancing and tramping leave sufficient marks; and the spot where the birds have met is like the ring of a circus after an equestrian performance. As it will of course be understood, it is the cocks only which are shot. And of these but a certain number; care always being taken to leave some of the old ones behind, to lead the young generation in the following season to the accustomed trysting place. And next year, in March, they are there on the very same spot as before."

The female—or, as she is commonly known, the Gray Hen—does not make much of a nest, and lays from eight to twelve eggs. The chicks make their appearance in three weeks' time, and leave the nest to follow their mother, who leads them to new fields, and gathers them under her wings wherever night overtakes them. She roosts upon the ground, and does not perch until the young are sufficiently strong to accompany her. During the period of incubation the male remains in the neighborhood, keeping vigilant watch over his family, and shows considerable skill in decoying any intruder from the vicinity of the brood. Late in the fall the males associate together in considerable numbers, and live peaceably with each other until the spring, when they again separate to seek the hens.

I have included as a synonym of this species the Tetrao Derbianus of Gould, which appears to be only an old Black Cock, with the tail feathers slightly elongated. This character I have observed in several examples, coming from various localities, and is hardly sufficient to constitute a separate species.

The adult male is black, with the head, neck, and back glossed with deep steel-blue reflections. Wings brown; a conspicuous band of white crosses the secondaries, which are also tipped with the same. Primaries brown, outer edges mottled with yellowish brown, and having shafts of a brownish white. Tail black, much forked, with the four lateral feathers on either side elongated and curved outward. Under tail coverts white, some in the centre projecting beyond the tail. Flanks and breast brownish black. Bill black. The legs yellowish brown, mottled with black. Feet brown. Superciliary membrane blood red.

Female has head and neck rufous, barred with brownish black; lower part of back and upper tail coverts of a deeper red, similarly barred. Upper part of breast light red, crossed with curved bars of black, and each feather broadly tipped with white. Abdomen mottled with dark brown. Flanks same color as the back, and similarly barred. The wings are reddish brown, mottled and barred with black, feathers tipped with an angular white spot. Primaries dark brown, mottled on their outer webs with reddish; secondaries similar, but their edges more broadly mottled and their tips white. Tail forked and black, irregularly marked with red, tipped with white, broadest on the central feathers. Under coverts white, sometimes with patches of brown or light red in the centre toward the end. The tarsi are covered with grayish white feathers mottled with brownish. Feet brown. Bill black.

COCK OF THE PLAINS.

CENTROCERCUS UROPHASIANUS.

CENTROCERCUS UROPHASIANUS. Swain.

COCK-OF-THE-PLAINS. SAGE COCK.

TETRAO UROPHASIANUS. Bon., Zool. Journ., vol. iii., Jan., 1828, p. 214.—Ib. Am. Ornith., vol. iii., 1830, pl. xxi., fig. 1.—Ib. Mon. Tetrao, Trans. Am. Phil. Soc., N. S., vol. iii., 1830, p. 390.—Ib. Geog. and Comp. List Birds, p. 44, No. 287.—Doug., Trans. Linn. Soc., vol. xvi., 1829, p. 133.—Nutt., Man., vol. i., 1832, p. 666.—Aud., Ornith. Biog., vol. iv., 1838, p. 503, pl. 371.—Ib. Syn., p. 205.—Ib. Birds of Amer., vol. v., 1842, p. 106, pl. 297.—Newb., Zool. Cal. and Or. Route, Rep. P. R. R. Surv., vol. vi., 1857, p. 95.—Gray, Gen. of Birds, vol. iii.

TETRAO (CENTROCERCUS) UROPHASIANUS. Swain and Rich., Faun. Bor. Amer., vol. ii., 1831, p. 358, pl. lviii.

CENTROCERCUS UROPHASIANUS. Baird, U. S. P. R. R. Exp. Exped., vol. iv., p. 624.—Gray, Cat. Birds, Brit. Mus, part iii., p. 48, 1844. —Cooper and Suckl., Nat. Hist. Wash. Territ., p. 222.—Jard., Game Birds, Nat. Libr. Birds, vol. iv., p. 140, pl. xvii.—Elliot, Proceed. Acad. Nat. Scien. (1864) p.

TETRAO PHASIANELLUS. Ord. Guth. Geog., 2d Am. edit., 1815, p. 317.

COCK-OF-THE-PLAINS. Lewis and Clark, vol. ii., p. 180, sp. 2.

This splendid bird, for its great size, stands pre-eminently in the front rank of the American Grouse, and is only exceeded in that particular, among all the members of this family, by the stately European Cock-of-the-Woods (Tetrao Urogallus), and its near ally (T. Urogalloides). The Sage Cock is never observed in the eastern portion of our continent, but dwells on the vast plains which lie on both sides of the Rocky Mountains, and wherever, on those almost endless tracts, the Sage Bush (Artemisia Tridentata) grows, there the Cock-of-the-Plains abounds.

The flight of this species is strong, and, at times, well sustained; it rises with the loud whir-r-r peculiar to this class of birds, and progresses by alternate flapping and sailing, generally in a straight line, until hidden by a hill or lost to the eye in the far distance. The courting season commences in the early spring, generally March or beginning of April. At such times, about sunrise, the male, perched upon some hillock, lowers his wings until the primaries rest upon the ground, spreads out his tail like a fan, and with the gular sacks inflated to a prodigious size, and head drawn back, he struts up and down before the admiring gaze of the assembled hens; then lowering his head until it is on a level with his body, he exhausts the air contained in the sacks, producing a loud grating noise resembling Anrr-hurr-r-r-hoo, ending in a "deep, hollow tone, not unlike the sound produced by blowing into a large reed." It is in this position I have endeavored to represent the male in the plate. The nest, formed of twigs and grass, is always placed upon the ground, near the bank of some stream, or sheltered by low bushes. The hen lays about fifteen or sixteen eggs, of a dark brown color, spotted on the larger end with chocolate. In about three weeks the chicks appear, and, like all of this family, run as soon as they are hatched, deserting the vicinity of the nest in a few hours.

During the summer and autumn these Grouse go in small flocks, sometimes only in pairs; but in the winter and spring they congregate in immense packs, to the number of several hundreds, and roam over the prairies in quest of subsistence. Their food consists chiefly of the leaves of the Artemisia, which, being very bitter, renders their flesh strong, and at times utterly unfit to eat, thus often depriving some hungry traveller on the plains of what promised him a delicious and savory meal. In the autumn, according to Nuttall, they frequent the streams of the Columbia River, where they feed on the Pulpy-leaved-Thorn; at which time they are considered good food by the natives, who take great quantities of them in nets.

Dr. Buckley, in his "Natural History of Washington Territory," speaking of this bird, says : " I have dissected these Grouse in situations where there was abundance of grass seeds, wild grain, grasshoppers, and other kinds of food that a person would imagine would be readily eaten by them, yet I have failed to obtain a single particle of any other article of food in their fall stomachs than the leaves of the Artemisia. This food most either be highly preferred, or else be essential to their existence. They seem to have the faculty of doing a long time without water, as I have found them in habitually dry, desert situations, during severe droughts, a long distance from water. I have found this bird most abundant on the southern slope of the Blue Mountains in the vicinity of Powder River. Here there are immense desert Sage plains, well adapted to the species in every respect. The bird hides well, and lies close, frequently allowing a man's approach to within a few feet."

With the following very interesting account of the Sage Cock, I close my article on this species. It is taken from the report upon the zoology of the route for a railroad to the Pacific Ocean by Dr. J. S. Newberry : "This is the largest of the American Grouse, the male sometimes weighing from five to six pounds. It is when in full plumage rather a handsome bird, at least decidedly better looking than any figure yet given of it. The female is smaller than the male, and of a monotonous sober brown; but the male, brown above, is handsomely marked with black and white on the neck, breast, and wings, and has a distinctive character in the spaces of bare orange-colored skin which occupy the sides of the neck. These spaces are usually concealed by the feathers, but are susceptible of inflation to a great size, and, when strutting in parade before the females, the neck is puffed out like that of the pouter pigeon. This bird does not inhabit the valleys of California, but belongs to the fauna of the interior basin, or, more probably, to the Rocky Mountain fauna—that of the dry, desert country lying on both flanks of the Rocky Mountain chain. We first met with it high up on Pit River, at the point where we left it, and crossed over to the lakes. Coming into camp at evening, I had been attracted by a white chalk-like bluff, some two miles to the right of our trail, which I visited and examined. Near it was a warm spring, which came out of the hillside, and spreading over the prairie, kept a few acres green and fresh, strongly contrasting with the universal brown of the landscape. In this little oasis, I found some, to me, now flowers, many reptiles, and a considerable number of Sharp-tailed Grouse, of which I killed several; the whole presenting attractions sufficiently strong—as we were to remain encamped one day—to take me over there early next morning. I had filled my plant case with flowers, had obtained frogs and snakes and chalky, infusorial earth enough to load down the boy who accompanied me, and had enjoyed a fine morning's sport, dropping as many Grouse on the prairies as we could conveniently carry. Following up the little stream toward the spring on the hillside, a dry, broken surface, with patches of 'sage bushes' (Artemisia Tridentata), I was suddenly startled by a great flutter and rush, and a dark bird, that appeared to me as large as a turkey, rose from the ground near me, and uttering a hoarse kik, kik, flew off with an irregular, but a remarkably well-sustained flight. I was just then stooping to drink from the little stream, and quite unprepared for game of any kind, least of all for such a bird, evidently a Grouse, but so big and black, so far exceeding all reasonable dimensions, that I did not think of shooting him, but stood with open eyes, and, doubtless, open mouth, eagerly watching his flight to mark him down. But stop he did not, so long as I could see him, now flapping, now sailing, he kept on his course, till he disappeared behind a hill a mile away. I was of course greatly chagrined by his escape; but knowing that, given one Grouse, it is usually not difficult to find another, I commenced looking about for the mate of the one I had lost. My search was not a long one; almost immediately she rose from under a sage bush, with a noise like a whirlwind, not to fly a mile before stopping to look around, as the cock had done, but, by a fortunate shot, falling helpless to the ground. No deerstalker ever felt more triumphant enthusiasm while standing over the prostrate body of a buck, or fisherman, when the silvery sides of a salmon sparkled in his landing net, than I felt, as I picked up this great and, to me, unknown bird. I afterward ranged the hillsides for hours, with more or less success, waging a war on these birds, which I found to be quite abundant, but very strong-winged and difficult to kill. I repeatedly flushed them not more than ten yards from me, and, as they rose, poured my whole charge right and left into them, knocking out feathers, perhaps, but not killing the bird, which, in defiance of all my hopes and expectations, would carry off my shot to such a distance that I could not follow him, even did I know he would never rise again. Here as elsewhere I found these birds confined to the vicinity of the 'sage bushes,' from under which they are usually sprung.

"A few days later, on the shores of Wright and Rhett lakes, we found them very abundant, and killed all we cared to. A very fine male which I killed there was passed by nearly the whole party within thirty feet in open ground. I noticed him perhaps as soon as he saw us, and waited to watch his movements. As the train approached, he sunk down on the ground, depressing his head, and lying as motionless as a stick or root, which he greatly resembled. After the party had passed, I moved toward him, when he depressed his head till it rested on the ground, and evidently made himself as small as possible. He did not move till I had approached to within fifteen feet of him, when he arose and I shot him. He was in fine plumage, and weighed over five pounds. We continued to meet with the Sage Hen, whenever we crossed sage plains, till we reached the Columbia. To the westward of the Cascade range this bird probably does not exist, as all its habits and preferences seem to fit it for the occupancy of the sterile and anhydrous regions of the central desert. Its flesh is dark and, particularly in old birds, highly flavored with wormwood, which to most persons is no proof of excellence. The young bird, if parboiled and stewed, is very good; but, as a whole, this is inferior for the table to any other species of American Grouse."

Among the specimens before me, is a very curious hybrid, between this species and Pedioecetes Columbianus. It was obtained by Mr. John Pearson, on the military road from the Walla Walla River to Fort Benton, and is marked on its label as No. 17,606 of the Smithsonian Institution collection. It is about the size of the Sharp-tail Grouse, but has the characteristic markings of the Sage Hen upon its head, neck, wings, and tail. The range of this species seems to be restricted to the desert plains which extend on both sides of the Rocky Mountains, and these birds are always more abundant wherever the Artemisia grows. The male may be described as follows: General color of back, light brown, each feather mottled and crossed irregularly with black and dark brown, and having also three bars of yellowish white, one near the tip, the other two higher up, equidistant from each other. The first of these is often almost obsolete. Some of the feathers in the centre of the back have broad bars of black, which cross and include the shaft, appearing like blotches upon the lighter ground color. This confused irregular marking extends throughout the upper tail coverts, and includes the two centre tail feathers. The tail is cuneate, longer than the wings, composed of twenty feathers, acute and graduated, and with the exception of the two centre ones, is of a dark brown color, crossed with irregular yellowish white lines, becoming fewer and at greater distances apart, upon the outer feathers. Upper part of head and neck crossed with zig-zag black and dark brown lines on a white ground in a very irregular manner. The wings are of a lighter brown than the back, crossed similarly with black, but having the shafts of the feathers all white, making them very conspicuous. The primaries are a dark brown, lighter on their outer webs, with dark brown shafts. The throat and under part of the neck is black interspersed with white lines and spots. A white band crosses the lower part of the neck, and extends over the sides, covering the position of the gular sacks. The feathers on this portion, especially those on the side, are very rigid, overlapping each other like scales, and in some specimens crackle like parchment when the hand is passed over them. The upper part of the breast is white, with the shafts black and stiff. The entire under parts, from the breast, are black, the under tail coverts black tipped with white. The black of the belly has a border of white blotched with black, while the flanks are mottled like the back. The feathers of the thighs and tarsi are light brown, mottled with a darker brown. The bill is thick and strong, black, with the nasal fossa extending nearly two thirds its length. The female resembles the male, but is smaller, and is without the gular sacks. The black of the lower parts is not so extensive, neither are the stiffened shafts of the neck feathers so conspicuous, while the bars and mottling of the upper parts is much greater.

The plate represents a male and female about three fourths the natural size. Different specimens, particularly among the males, vary considerably in size, some being nearly half as large again as the one represented.

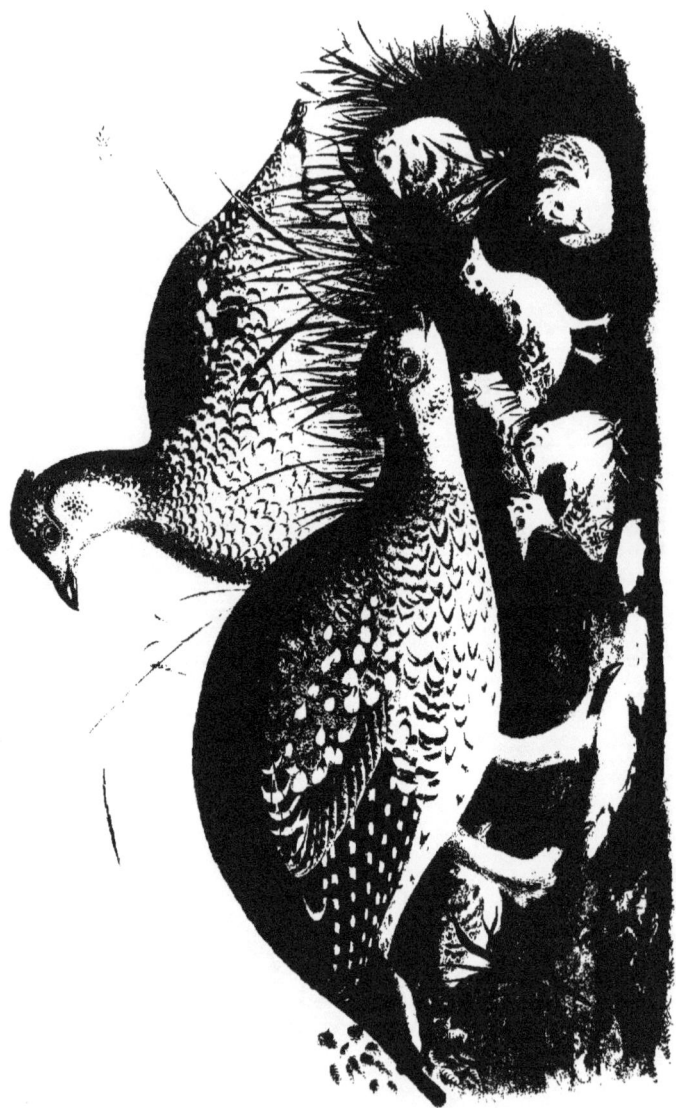

PEDIAECAETES COLUMBIANUS. Elliot.

SHARP-TAILED GROUSE.

TETRAO PHASIANELLUS. Ord. Guthr. Geog., 2d Amer. edit., ii., 1815, p. 317.—Nutt., Man., vol. i., 1832, p. 669.—Aud., Orn. Biog., vol. iv.,
1838, p. 509, pl. 382.—Ib. Syn., 1839, p. 205.—Ib. Birds of Amer., vol. v., 1842, p. 110, pl. 298.—Newb., Zool. Cal. and Or. Route.
Rep. P. R. R. Surv., vol. vi., 1857, p. 94.—Bon. Syn., 1828, p. 127.—Coop. and Suckl., Nat. Hist. Wash. Territ., p. 223.—Bon. Am.
Ornith., vol. iii., p. 44, plate.
PHASIANUS COLUMBIANUS. Ord. Guth. Geog., 2d Amer. edit., 1815, vol. ii., p. 317.
TETRAO UROPHASIANELLUS. Doug., Trans. Linn. Soc., vol. xvi., 1829, p. 196.
PEDIAECAETES COLUMBIANUS. Elliot, Proc. Acad. Nat. Sciences (1863), p.   , and 1864, p.
PEDIAECAETES PHASIANELLUS. Baird, U. S. Ex. Exp. P. R. R., vol. ix.

This fine bird, often confounded with the well-known Pinnated Grouse or Prairie Chicken (*Cupidonia Cupido*), dwells on the plains bordering the Mississippi and Missouri rivers, and so on westward across the continent. It has never, I believe, been obtained to the east of the Mississippi, but supplies, on the vast western plains, the place of its near ally. In their habits these two species somewhat resemble each other, but the Sharp-tail seems to be destitute of the gular sacks so prominent in the other during the spring. They congregate in flocks, sometimes of many hundreds, and as they lie close, and fly only a short distance on being disturbed, afford very good sport to the gunner. They rise with the whirring noise, caused by the rapid beating of the wings, common to this family, and as they commence their flight, utter a clucking sound often repeated. They fly generally straight and rather swift, but in the fall are easily brought down by a cool sportsman. Their flesh resembles that of the "prairie chicken;" in fact, I have been unable to distinguish the one from the other, when both have been served up together. The present species is never found on the high lands or in the forests, but is only to be procured upon the prairies, which are alone its natural home.

Dr. Suckley says of this Grouse, that "We first noticed the species in Nebraska, near Fort Union, at the mouth of the Yellowstone River. From that point to the Cascade Mountains of Oregon and Washington Territories, the species is exceedingly abundant, wherever there is open country and sufficiency of food. In certain places they are in great numbers in the autumn, congregating in large flocks, especially in the vicinity of patches of wild rye, and more recently near settlements where there are wheat stubbles. They resemble the Pinnated Grouse in habits, and are good both for table and for sport. In places where they are numerous, they may frequently be found on cold mornings in the autumn or early winter, perched on fences or leafless trees, sunning themselves in the early sunlight. At Fort Dalles, on the 1st of April, 1855, a young bird scarcely two days old was brought to me. This early incubation would lead us to suspect that the species in favorable situations has two or more broods during the season. The young bird above mentioned was confided to the motherly care of a hen with a young brood of chickens, but the young Grouse, not understanding the kindness of its foster-parent, ran and hid itself as soon as possible, and probably perished that very night from cold."

This species has been considered by ornithologists generally as the same as Linnæus's *Tetrao Phasianellus*, and is mentioned, in the various works and papers devoted to this science, by that specific appellation. A number of specimens of Sharp-Tailed Grouse having arrived at the Smithsonian Institution from Arctic America, it was discovered on examination that they were the species described by the great Swede, and that our familiar bird was probably unknown to him. The points of difference will be fully described in my article on the *Pediaecaetes Phasianellus*.

The plate represents the male, female, and young of the natural size. The latter I believe have never before been figured; and I am indebted to my friend W. J. Hays, Esq., well known as an artist unequalled in this country for his pictures of animal life, for the opportunity of introducing them into my plate. These chicks were obtained by this gentleman during an excursion he made a short time since up the Missouri river, and are the only specimens that I am aware of, in any cabinet in this country. This species is sometimes brought to the markets in this city with the Pinnated Grouse, which are sent from the extreme West, when there is a long continuance of cold weather. It is not generally distinguished by the poultry vendors from the better-known grouse, although by some of them it goes by the name of white-breasted prairie chicken.

PEDIAECAETES COLUMBIANUS.

I have at various times obtained hybrids between this species and *Cupidonia Cupido*, and also between it and the more northern Sharp-tail, *Pedioecetes Phasianellus*. Some of the offspring of this species and the Prairie Chicken are very handsome birds, having a good deal of the pure white under parts of the Sharp-tail, but the upper part of the breast and the flanks are crossed with bars scolloped on the lower edge, instead of the single heart-shaped spots, making a very peculiar and striking effect. Of course these hybrids vary a good deal in their markings, accordingly as the Prairie Chicken or the present species predominates, for some incline to one species more than to the other. This species is distributed from the Mississippi, throughout the northern and western prairies, to Oregon and Washington Territories.

Head and throat brownish-yellow, the front, crown, occiput and cheeks irregularly marked with black or very dark brown; superciliary stripe whitish; back ferruginous brown, variously spotted with black or brownish yellow; wings brownish gray, with large spots of white on all the coverts; transverse bars on the secondaries, and the outer webs of the primaries, which are dark brown, spotted with the same; the tail feathers have the inner web white, outer, brownish gray, dotted with darker brown, the central feathers are elongated and of the same color as the back; under parts pure white, the feathers on the breast and flanks having a brown U-shaped mark. Bill black; feet brown. There is no difference in color of plumage between the sexes. The young have the upper parts a light brownish yellow, crossed irregularly with lines of blackish brown; wings pinkish white, barred with black. Entire under parts yellow, darker on the sides and upper part of breast. Thighs and tarsi same color as belly. Bill light yellow, with a central brown line on upper mandible.

PEDIAECAETES PHASIANELLUS. Elliot.

TETRAO PHASIANELLUS. Bon., Geog. and Comp. List Birds, p. 44.—Lath., Ind. Ornith., vol. ii.. p. 635.—Linn., Syst. Nat., vol. i., edit. 10th, p. 160 (1758).—Sab. Frank., 1st voy., p. 680.—Forst., Philos Trans., lxii., 1772, p. 394 and 405.

CENTROCERCUS PHASIANELLUS. G. R. Gray, Cat. B. Brit. Mus., Part III.—Bon., Compt. Rend., xiv., p. 428 (1857).

TETRAO (CENTROCERCUS) PHASIANELLUS. Swain, Faun. Bor. Amer., vol. ii., p. 361 (1831).

SHARP-TAIL GROUSE. Penn, Arct. Zool., vol. i., p. 357, No. 181.—Hearne's Journ., p. 408.

LONG-TAILED GROUSE. Edw., Birds, vol. iii., p. 117.

TETRAO UROGALLUS. Linn., Syst. Nat., vol. i., edit. 12th, p. 273.—Var. B.

PEDIOCAETES KENNICOTTI. Suckl., Proc. Acad. Nat. Scien. (1861).

PEDIOCAETES PHASIANELLUS. Elliot, Proc. Acad. Nat. Scien., p. 403 (1862 and 1864).

AW-KIS-COW. Cree Indians.

This species, heretofore confounded with the Sharp-tail Grouse inhabiting the western portions of the United States, is found in Arctic America, plentiful around Hudson's Bay, but never yet, I believe, has it been obtained within the limits of the Union. It is easily distinguishable from its near ally, its prevailing black and white colors forming a strong contrast to the brownish yellow of the P. Columbianus.

My friend Mr. Kennicott, well known for his successful labors in the various branches of natural history, amid the wilds of the frozen North, sent to the Smithsonian Institution many examples of this species, obtained by him in his last expedition. These were the first ever in the possession of any American ornithologist, and were named by Dr. Suckley in honor of the gentleman who procured them, as they were evidently very different from the bird commonly known as P. Phasianellus, so abundant in some portions of our Western prairies. But, on a more critical examination, it was found that this was the species to which Linnæus had long since given the name of *Phasianellus*, and, consequently, it of course took precedence over that of P. Kennicotti, which sank into a synonym.

The present bird resembles its relative in its habits, goes in flocks, and is destitute of any gular sack. It may be described as follows:

General color black. Top of head black, a few faint marks of rusty toward the occiput, sides of head black, the feathers tipped with white; those on the side and back of neck tipped with rusty; throat white, spotted with black. The back is also black, the feathers margined with rufous brown; the rump is lighter, caused by the feathers being tipped broadly with grayish; the elongated central feathers of the tail are jet black, irregularly crossed with yellowish white and gray. Wings blackish brown, with large white spots on all the coverts in addition to the rusty margins of the feathers; primaries blackish, with white marks on their outer webs. Tail sometimes grayish at the base, with white tips or pure white. Under parts pure white, with a black V-shaped mark, near the centre of the feathers on the breast and flanks, gradually growing smaller and fainter as they approach the abdomen and vent. The white feathers of the legs are hair-like, and extend over the toes quite to the nails. Bill black; feet dark brown.

The figures in the plate are of the size of life.

THE PINNATED GROUSE.

CUPIDONIA CUPIDO.

CUPIDONIA CUPIDO. BAIRD.

PINNATED GROUSE.

TETRAO CUPIDO. Linn., Syst. Nat., vol. i., 1766, p. 274.—Gmel., Syst. Nat., vol. i., p. 751.—Lath., Ind. Ornith., vol. ii., 1790.—Wils., Am. Ornith., vol. iii., 1811, p. 104, pl. xxvii.—Bon., Obs. Wil., 1825, No. 169.—Id., Mon. Tetrao, Am. Phil. Trans., vol. iii., 1830, p. 392.—Nutt., Man., vol. i., p. 662.—Aud., Ornith. Biog., vol. ii., 1634, p. 490, and 1830, p. 550, pl. 186.—Id., B. of Amer., vol. v., 1842, p. 93, pl. 296.—Dou., Geog. and Comp. List, p. 44, No. 285.—Id., Syn., 1828, p. 127.
BONASA CUPIDO. Steph. Shaw's Gen. Zool., vol. xi., p. 299.
CUPIDONIA AMERICANA. Reich., Av. Syst. Nat., 1850, p. xxix.—Bon., Comp. Rend., xlv., 1857, p. 428.
TETRAO CUPIDO. Gray, Gen. of B., vol. iii.
CUPIDONIA CUPIDO. Baird, U. S. P. R. R. Exp. and Surv., vol. ix., p. 638.—Elliot, Proc. Acad. N. S. (1864).
LE COQ DE BRUYÈRE À FRAISE. Cuv., Reg. Anim., vol. i., p. 449.

FORMERLY this valuable species was found in great numbers from the Atlantic coast to the Mississippi River, but now it has almost entirely disappeared from the eastern shore, and even in the West is becoming gradually scarcer every year.

Early in the spring, at break of day, the prairies of the West resound with the loud booming notes of the excited males, who, meeting, sometimes in large numbers as though by appointment, strut up and down, with their feathers ruffled, gular sacks extended, and the long tufts on the neck raised above their heads, forming a kind of crest; until, overcome with their pugnacious feelings, they fight furiously. These conflicts, although carried on with much earnestness, seldom result in any injury to the combatants, the weak birds giving up the strife from sheer exhaustion, leaving the others to seek the hens, which have probably been looking on from the neighboring bushes.

The hooting of the Pinnated Grouse is made by exhausting the air in the gular sack, in like manner as the Cock-of-the-Plains, and consists of three notes, which in the clear atmosphere of the prairies may be heard for nearly half a mile. The males alone have the power of producing these sounds, the females being destitute of the necessary apparatus. Should the air-sacks become punctured, the bird is unable to hoot any more, although he will go through all the motions requisite to produce the sounds.

The Prairie Chicken makes its nest generally in April, and places it near some tuft of long grass on the open prairie. It is carelessly formed of leaves and grass, and usually contains twelve eggs, and the young make their appearance in about three weeks after incubation commences. Only one brood is raised during the season, although, should the eggs be destroyed, the female will lay again.

This species carries itself very erect when upon the ground, but is not so graceful as the Ruffed Grouse. If startled, should the object of its fear not be very near, it endeavors to escape by running, until, having reached some tuft, or clod of earth, it suddenly squats close to the ground, and remains until flushed.

The Pinnated Grouse roost upon the ground, within a short distance of each other; and this habit is frequently taken advantage of by the trapper, who, having previously marked the spot, goes to it after nightfall with a net, and often succeeds in capturing the greater part of the flock at one haul. This practice, however, has the effect of causing the survivors to desert the place, and it is much to be regretted that such wholesale destruction of these birds cannot be prevented; but it would seem that it is likely to continue until the last Pinnated Grouse has been taken.

In order to cleanse their feathers from insects, or any substance that may cling to them, they are fond of dusting themselves in the roads or ploughed fields; and they may often be seen thus occupied, during fine days.

Their flight is strong and well sustained, sometimes rapid. They propel themselves by several beats repeated in quick succession, and then sail onward for some distance with the wings slightly bent downward, when the beats are again renewed.

In August and September these birds are very gentle, and, in these months, very many are shot, as they will lie well to a dog, and are easily approached; but in the fall, when the young are fully grown, they pack; that is, many families join together, sometimes to the number of several hundreds, and are then very wild, rising out of gunshot, and continuing their flight often for more than a mile. If followed immediately and again started, they will frequently, on alighting, spatter and lie close, when the sportsman is enabled to obtain many of them.

At this period of the year they are fond of frequenting the cornfields, to pick up the grain which may be on the ground, returning to the prairie toward evening to roost. It is very difficult to approach them when among the corn, as a person makes so much noise passing between the stalks, that the birds become alarmed, and take to flight often unobserved. Many an hour have I passed, toiling after them in those unfavorable localities, and considered myself fortunate if five or six birds were the result of the hunt.

In winter the Prairie Chickens perch upon the fences, and early in the morning the topmost rails for a long distance are often completely hidden by the multitude of Grouse which have settled on them. As soon as the sun is two or three hours high, they leave their perches to seek their food.

This species is capable of going for a considerable time without water, the districts they inhabit being generally dry : and they are accustomed to quench their thirst by picking off the drops of rain or dew that glisten upon the leaves and grass.

Their flesh, when young, is white, but in the adult quite dark ; and is generally much esteemed as an article of food.

Unlike the Ruffed Grouse, which is of an untamable disposition, the Prairie Hen is easily domesticated, and will breed in confinement. When this species takes flight, it is with much less whirring of the wings than is characteristic of other members of this family, and frequently, on rising, they utter a few distinct clucks.

The two sexes resemble each other closely in their plumage, the principal difference being that the male possesses the gular sacks, and tufts of lengthened feathers upon the sides of the neck. They may be described as follows:

General color of the upper parts brown, transversely barred with blackish brown : wings lighter brown ; primaries grayish brown, with spots of reddish yellow on the outer webs. Tail-feathers purplish brown, the two middle ones lighter and mottled with brownish black. Loral space and throat light buff. The long feathers of the neck are yellowish red, dark brown on the outer webs. Under parts white, marked with broad curved bands arranged in regular series, of a grayish brown; under-tail coverts white, crossed with brown and margined with black. Membrane over the eye, and gular sack, orange yellow. Bill dusky ; feet yellow. Feathers of the legs gray, minutely banded with yellowish brown.

The plate represents a male in the act of *hooting* to a female surrounded by her brood.

The figures are all life-size.

LAGOPUS ALBUS.

WILLOW PTARMIGAN.

TETRAO ALBUS. Gmel., Syst. Nat., vol. i., 1788, p. 750.—Lath., Ind. Ornith., vol. ii., p. 639.

TETRAO SALICETI. Temm., Man. d'Ornith., p. 471, vol. i.—Sab., Appen. Frank. Narr., p. 681.—Rich., Appen. Parry, 2d Voy., p. 347.—Aud., Ornith. Biog., vol. ii., 1834, p. 526, pl. 191.

TETRAO (LAGOPUS) ALBUS. Nutt., Man. Ornith., vol. i., 2d edit., 1840, p. 816.

TETRAO (LAGOPUS) SALICETI. Swain, F. Bor. Amer., vol. ii., 1831, p. 351.—Ross, Arct. Exp., p. 28.

LAGOPÈDE DE LA BAIE D'HUDSON. Buff., vol. iii., p. 310.—Id., Ois., vol. ii., p. 276.—Cuv., Reg. Anim., p. 449.

TETRAO LAGOPUS. Forst., Phil. Trans., lxii., 1772, p. 390.

WHITE GROUSE. Pennant, Arctic Zool., vol. i., p. 360.—Lath., Syn., vol. iv., p. 743.

WHITE PARTRIDGE. Edwards, Birds, pl. 72, male in change.

WILLOW PARTRIDGE. Hearne, Journ., p. 411.

PERDIX DES SAULES. Hearne, Voy. à l'Océan du Nord, p. 338, edit. in 4°.

TETRAO DES SAULES, OU MUET. Temm., Pig. et Gall., vol. iii., p. 208, t. Anat., 11, f. 1, 2, and 3.

TETRAO LAPPONICUS. Gmel., Syst. Nat., vol. i., p. 751, sp. 25.—Lath., Ind. Ornith., vol. ii., p. 640, sp. 12.

WEISSE WALDHUHN. Bechst., Naturg. Deut., vol. iii., p. 1253.

TETRAO HEHUSAK. Temm., Pig. et Gall., vol. iii., p. 225.

REHUSAK GROUSE. Lath., Syn. Suppl., vol. i., p. 216.—Penn. Arct. Zool., vol. ii., p. 316.

LAGOPUS ALBUS. Ben., Am. Phil. Trans., vol. iii., N. 3., p. 393, sp. 313.—And., Syn., 1839, p. 207.—Id., Birds of Amer., vol. v., 1842, p. 114, pl. 299.—Gray, Gen. of Birds, vol. iii.—Baird, U. S. P. R. R. Exp. and Surv., vol. ix., p. 633.—Boie, Isis, 1822, p. 558.—G. R. Gray, Cat. B. Brit. Mus., Pt. III., p. 47 (1844).—Bon., Geog. and Comp. List Birds, p. 44, No. 288.

LAGOPUS SALICETI. Gould, Birds of Europe, pl.

LAGOPUS SUBALPINUS. Nils., Orn. Suec., vol. i., p. 307, sp. 139.

LAGOPUS BRACHYDACTYLUS. Temm., Man. d'Ornith., vol. iii., p. 328.—Gould, B. of Eur., pl. 256.—Gray, Gen. of B., vol. iii.—Bon., Geog. and Comp. List B., p. 44, No. 300.

THE Willow Grouse is an inhabitant of the northern portions of both hemispheres, but is rarely seen within the limits of the United States. Audubon mentions that he had seen the skins of several that were shot near Lake Michigan, and also states that he felt assured it existed in the State of Maine. Although I do not think that the Willow Ptarmigan is an habitual resident of any part of the Union, yet in very cold winters it has frequently migrated southward, and been taken within our borders.

This species is monogamous, and the male remains in the vicinity of the nest while the female is sitting, and afterward accompanies the brood until they are nearly full grown, evincing much affection and tenderness toward them. The female constructs her nest of twigs and mosses, and lays about a dozen eggs of a rufous color, thickly spotted with reddish brown. But one brood is raised during the season, and the young are at first covered with a yellowish down.

Audubon, speaking of the affection possessed by the adults for their offspring, states "that when a covey happened to come in our way, the parents would fly directly toward us with so much boldness, that some were actually killed on the wing with the rods of our guns, as they flew about in the agonies of rage and despair, with all their feathers raised and ruffled. In the mean time, the little ones dispersed and made off through the deep moss and tangled creeping plants with great rapidity, squatting and keeping close to the ground, when it became extremely difficult to find them."

The flight of the Willow Ptarmigan is regular and swift, sometimes protracted to a great distance; and on rising, they utter a cluck several times repeated. In winter they associate together in large flocks, and obtain their subsistence mainly by feeding upon the lichens and moss, to reach which they are obliged to scratch away the snow.

LAGOPUS ALBUS.

The principal food of this species consists of the leaves of plants, and sprouts of several kinds of willow, berries, &c. Wheelwright, in his account of this Ptarmigan, as quoted by Bree in his "Birds of Europe," says that "the Willow Grouse, in summer, is usually found in valleys, mostly by the side of the little books or mountain streams which run among the bushes and thickets. You always find them in pairs or families, with the male and female together. You not only find them, according to Nilsson, in the interior of the country, but even on the coasts and islands. They crouch among the dwarf birch, willow, or heather, and rarely rise until you nearly tread on them.

"Sometimes, however, they rise very wild, and in the spring and autumn appear most shy. They almost always are on the ground, and very rarely perch in a tree; but although I have myself seen, on more than one occasion, the Willow Grouse, when frightened, perch in the birch trees, it is so rare an occurrence that many deny it. Their flight appears to me exactly to resemble that of the Red Grouse, and as they fly they utter a loud cackle which much resembles 'cvveckackackh.'"

In the shape and size of the bills of these birds the most astonishing differences exist, and I have never been able to find two exactly alike. They range all the way from the robust and powerful to those almost as small and delicate as characterize the Lagopus Mutus.

This peculiarity is not confined to particular localities, else it might almost be considered of specific value; but members of the same flock will differ in this respect as much among themselves, as though they were indeed of separate origin.

With specimens before me from Lapland, Sweden, Denmark, Finland, Russia, and throughout the northern portion of the American continent, I find this variation in the bill common to all, and do not consider it as indicative of any specific distinctness, as these specimens which are in the summer dress present a general similarity of coloring in their plumage.

Baird, in his valuable work on "North American Birds," in the article on this species, speaks of some specimens in his possession from Labrador and Newfoundland, which "appear to have decidedly broader, stouter, and more convex bills than those from Hudson's Bay and more northern countries," and says that it is possible there may be two species.

Among the large number of Willow Grouse obtained by Mr. Kennicott, I find specimens from Great Slave Lake, Mackenzie River, and Fort George, as well as others brought from Lapland and Sweden, which have as large bills as any from Newfoundland: while from these same localities, and out of the same flocks, are other examples with much smaller and more feeble bills. It would therefore seem to be the reasonable conclusion, that, unless the summer dress should be very unlike the typical style of L. Albus, this variation in the bill must be deemed as of no particular importance in deciding the species to which the bird may belong, but merely one of those unaccountable freaks of nature occasionally met with. Thus far I have not seen any specimen of Ptarmigan of the Albus style, in its summer dress, which presented undoubted evidences of belonging to a different species.

All Ptarmigan vary so much from each other, that, in order to determine a good species, many adult specimens in summer plumage must be available; for I do not consider it at all likely that any species of Ptarmigan, established solely upon a skin of the bird when in its winter plumage, would stand the test of future research, as it could present but few, if any, reliable differences to distinguish it from others.

Some specimens, in winter, have the bill nearly covered with feathers, giving to it the appearance of being quite small, whereas in summer it would be the reverse; and the feathers on the legs and feet grow so long and thick as to cause the latter to seem shortened.

I have included among the synonyms of this species the Lagopus Brachydactylus of Temminck, as I cannot perceive any satisfactory differences given in his description to constitute his specimen as distinct. The bill, nearly hidden by feathers, is characteristic of all Ptarmigan in the winter dress, as is, also, having the legs and feet densely covered.

As these birds vary so greatly in the color of their plumage, it is not surprising that some should have the shafts of the primaries white; but this would not be a reliable character to establish a species upon. I have seen specimens of L. Albus in winter dress, which had the shafts of some of the primaries nearly white, while the rest were dark brown.

From the measurements given by Temminck, although they are rather less than is usual, this bird would seem to belong to the Lagopus Albus; but as it has no black stripe through the eye, it may possibly be a female of the L. Mutus.

Amid a number of Ptarmigan sent to me by Mr. Gould from London, for my inspection, and to whom I take this opportunity of expressing my thanks for his kindness, was a specimen from Arctic America, answering to the description of Brachydactylus, which, excepting the white shafts of the primaries, agreed with many examples in my possession from Great Slave Lake. As none of these last, in the summer dress, present any material differences from the L. Albus, I cannot but consider them as identical.

Another specimen in Mr. Gould's series was a hybrid between this species and the Canace Canadensis. It was in summer dress, and had the slender nails and structure of feathers of the Spruce Grouse; while a number of white feathers, showing an evident inclination to change in winter, betokened the Ptarmigan blood.

The Willow Ptarmigan dwell chiefly upon the plains, and in this respect differ from the Lagopus Mutus, which makes its home upon the mountains, near the line of perpetual snow. Richardson states that, "like most other birds that summer within the Arctic circle, they are more in motion in the milder light of night than in the broad glare of day."

This species has a very wide distribution, as it is found in the high latitudes of both the Old and New Worlds, being abundant in Sweden, Norway, Lapland, Russia, Siberia, Greenland, and throughout the Arctic regions of North America. It is not found in the British Islands.

In summer, the adult male has the head, neck, and breast chestnut, sometimes very dark, nearly black on the breast; barred on the top of the head and back of the neck with black; chin sometimes black, with a white spot on each side. Rest of upper parts black, transversely barred with reddish yellow. Tail black, tipped with white, the two centre feathers marked like the back. Wings, abdomen, thighs, and legs pure white. Under tail coverts brown, barred with black. Nails long, brown, and flat beneath.

The female, in summer, is rusty yellow in color, thickly barred and blotched with dark brown and black. This yellow hue extends throughout the lower parts, sometimes white feathers appearing about the abdomen; the flanks are barred with blackish brown. Wings, legs, and feet pure white. Tail black, tipped with white; upper coverts mottled like the back; lower coverts yellowish, barred with dark brown. Bill black. Nails similar to those of the male.

WILLOW GROUSE.

LAGOPUS ALBUS. — (Willow Ptarmigan) — Adult Female.

In both sexes there is a vermilion membrane over each eye, larger and more conspicuous in the male.

In winter the plumage is pure white, with the exception of the tail, which is black tipped with white. The feathers of the legs and feet grow very long and thick at this season, covering the entire foot, sometimes even the nails. These last are very long, thin, concave beneath, and white. They are shed generally in July.

The young have the head above dark brownish red, with a black spot on the crown, and a blackish-brown stripe on the back of the neck. Upper part of the body speckled with rufous brown and black, with white spots on the wings. Under parts yellowish. Tail yellowish brown, barred irregularly with blackish brown.

This species seems to be constantly moulting, as it is very rare to find a specimen that has not any new feathers upon some part of its body. The change from the summer to the winter dress commences generally toward the latter part of September, and the birds present a very pretty speckled appearance with the white feathers intermingled with the colored ones.

The spring moult commences first with the males, the colored feathers appearing upon the head and neck, which, from their red hue, when contrasted with the white bodies, have a very peculiar effect when viewed from a distance. The mottled feathers of the back next appear, and lastly those of the breast. The feathers near the nostrils seem to remain white the longest, as I have frequently observed this in specimens which otherwise were in complete summer plumage. The back figure in the plate shows this peculiarity.

The female does not commence to change until several days after the male, but proceeds more rapidly with the moult, and assumes her full livery nearly at the same time as the other sex.

I have considered it best to give two plates of this species, so as to show it in both the summer and winter dress.

The first represents two adult males, and a female surrounded by her brood; the other, one bird in process of change, and one in its pure white livery. All the figures are life-size.

LAGOPUS SCOTICUS.

RED GROUSE.

TETRAO SCOTICUS. Lath., Ind. Ornith., vol. ii., p. 641, sp. 15.—Selby, Brit. Ornith., pl. xix.—Lewin's Brit. Birds, t. 188.—Mont., Ornith. Dict. and Supp., vol. i.—Bewick's Brit. Birds, vol. i., p. 351.—Graves, Brit. Ornith., vol. ii.—Jenyns, Man. B. Vert. Anim., p. 170.
LA GELINOTE D'ÉCOSSE. Briss., i, p. 199, 5, t. 22, f. 1.
TETRAO LAGOPUS. Gmel., Syst. Nat., vol. i., p. 750. Var. 7 and 8.
POULE DE MARAIS GROCE. Cuv., Reg. Anim., vol. i., p. 460.
TETRAS ROUGE. Temm., Man. d'Ornith., vol. i., p. 465.
LAGOPUS ALTERA. Albin., vol. i., p. 386.
RED GROUSE. Penn., Brit. Zool., 1, No. 94, t. 43.—Lath., Syn., 4, p. 746, and Supp., p. 216.—Morris, Nat. Hist. Brit. Birds, vol. ii., p. 342, pl. 171.—Yarr., Brit. Birds, 2 edit., vol. ii., p. 351.
RED GAME, MOOR COCK. Rail., Syn., p. 54.—Will. (Ang.), p. 177.
TETRAO SALICETI SCOTICUS. Schleg., Rev. Crit. des Ois. d'Eur., p. 76.
OREIAS SCOTICUS. Kaup., Naturl. Syst., p. 177.
BONASA SCOTICUS. Briss., Ornith., vol. i., p. 190.
LAGOPUS SCOTICUS. Gould, Birds of Eur., pl. 252.—Gray, Gen. of Birds, vol. iii.—MacGill, Brit. Birds, vol. i., p. 169.—Leach, Syst. Cat. Mam. and Birds Brit. Mus., p. 27.—Vieill., Nouv. Dict. d'Hist. Nat., xviii., p. 206.—Steph., Gen. Zool., vol. xi., p. 293.—Flem., Brit. Zool. p. 43.—Eyton, Cat. Brit. Birds, p. 30.—Don., Geog. & Comp. List of Birds, p. 44.—Gray, Cat. Birds Brit. Mus., Part III., p. 142 (1850).—Bon., Rev. Ornith. Eur., p. 174.

THE Red Grouse, Moor Cock, Red Game, Scotch Grouse, by all of which appellations this species is known, is an inhabitant of the British Isles. It is found in considerable numbers in various parts of England, Wales, and Ireland, but nowhere in such abundance as among the Highlands of Scotland. They love the moors covered with the purple heather, and resort chiefly to those tracts lying between the lofty haunts of the Ptarmigan and the lower, more wooded lands, where the Black Grouse are found.

This species is monogamous, the female laying from eight to twelve eggs; and during incubation, which is performed by her alone, the male keeps a vigilant watch at a short distance, waiting the appearance of the young to assist his mate in bringing up their family.

The Carrion Crow is said to hunt the moors at this period for the nests, and makes great havoc among their contents, whenever successful in finding them, even persecuting and sometimes killing the young chicks.

The Red Grouse has many enemies, from whose daring attacks it often has no little difficulty in escaping. Among the principal depredators, may be mentioned the Golden Eagle, Peregrine Falcon, and Common Buzzard.

Various kinds of berries and grasses constitute the food of the Moor Cock, and it is also very partial to oats and corn, and will feed to excess upon these whenever they are grown near its haunts. The fresh twigs of the heath also are eaten by it, the tips being broken off in small pieces.

The season for "Grouse shooting" commences about the middle of August, and the number of birds which are killed annually is so large, that one would suppose the species must become extinct in a very few years; yet they appear to suffer very little diminution.

To see the Scotch Grouse in his native home, one must go to the Highlands, and traverse the moors, which stretch away for miles on every side; and there, in the early morning, the male will be heard, perched upon some hillock, uttering his challenge-cry.

The following lines, known as the "Grouse-Shooter's Call," well describe the scene:

"Come, where the heather bell,
Child of the Highland dell,
Breathes its coy fragrance o'er moorland and lea;
Gayly the fountain shine,
Leaps from the mountain green—
Come to our Highland home, blithesome and free!

"See! through the gloaming
The young morn is coming,
Like a bridal veil round her the silver mist curled;
Deep as the ruby's rays,
Bright as the sapphire's blaze,
The banner of day in the east is unfurled.

"The Red Grouse is waitcaring
Down from his golden wing,
Gemmed with the radiance that heralds the day;

Peace in our Highland vales,
Health in our mountain gales—
Who would not hie to the mountains away?

"Far from the haunts of man
Mark the gray Ptarmigan,
Seek the lone Moor Cock, the pride of our dells;
Birds of the wilderness
Here in their resting place,
'Mid the lovers heath where the mountain roe dwells.

"Come, then; the heather bloom
Woos with its wild perfume,
Fragrant and blithesome thy welcome shall be;
Gayly the fountain shine
Leaps from the mountain green—
Come to our home of the moorland and lea!"

The voice of the Red Grouse is loud; and few, who have ever crossed the moors, but have heard his cackling note, resembling *kok, kok,* quickly repeated. This sound is chiefly uttered when the birds are disturbed; at other times they emit a loud cry which, as stated by MacGillivray, is easily syllabled into *go, go, go, go, go-back, go-back;* although the Celts, naturally imagining the Moor Cock to speak Gaelic, interpret it as signifying *co, co, co, co, mo-chblaidh, mo-chblaidh*—that is, *who, who, who,* (goes there?) *my sword, my sword!*

Toward winter, this species associate together in large flocks, and do not separate again until the following spring. When gathered together in such numbers, they are said to "pack," and are very shy, and difficult of approach, keeping at all times a vigilant watch on those who intrude upon their domains.

The coloring of the plumage of this species varies very much in different individuals, the majority appearing like those represented in the plate, which may be considered as the typical style; but I have seen specimens which had the entire breast almost black, without any mottling whatever. One kindly lent to me by Mr. Gould, of London, was of this description. Mr. Selby states, that those bred upon the moors of Blanchland, in the County of Durham, are of a cream color, or light gray, spotted more or less with dark brown or black. Sir William Jardine possesses a "Grouse shot on the moors of Galloway, where the ground color is nearly yellowish white, and all the dark markings are represented by pale reddish brown; the quills are dirty white. In some instances the plumage takes an opposite shade, and is remarkable for its deep tint, and the almost entire absence of markings. The whole, or part of the quills, are often found white."

Many genera have been assigned to this bird, and its specific names are very numerous; the doubts regarding it seeming to arise chiefly from the difficulty of defining its proper position, as to whether it should be included among the Grouse or Ptarmigan. It is undoubtedly nearest allied to the latter (the fact of its not turning white in winter being the strongest point of difference), and, like the Ptarmigan, it is feathered to the end of the toes, which circumstance is never observed in the true Grouse; although, when these inhabit very cold countries, the feathers of the tarsi grow very long, even covering the feet, and in this way protect the toes from the piercing air. I have noticed this more particularly in specimens of the Podinocaretes Phasianellus from Hudson's Bay. The fact of its varying so much in the color of its plumage, as cited above, is another evidence of its close affinity to the Lagopidae, of which genus it is almost impossible to find any two members exactly alike.

The species to which the Lagopus Scoticus approaches most closely, is the Lagopus Albus of Gmelin; so near it, indeed, that it may almost be considered as but an insular variety; and specimens of L. Albus resemble each other in color of plumage much more than do examples of any other species of Ptarmigan. Yet it would be unwise, perhaps, to consider these two as only one, for each present certain characters not observable in the other, sufficient to distinguish them easily, at all times. One might, without difficulty, speculate upon the origin of the Red Grouse, as to whether it is but an offshoot of the Willow Grouse, or whether, if transplanted to a more severe climate, where the winter lasts the greater portion of the year, it too might not, after a while, also turn white as the summer disappeared; these, after all, although argued with ever so much ability, would be but theories, and it is better to consider the *facts* as they present themselves to us at the present day, and draw our deductions from them, than to grope in the dim past, with but very insufficient guides to lead us to the truth, far which we all are striving. Without doubt, the white garb with which nature has clothed the Ptarmigan during the severer portions of the year, is in its very color an additional protection from the cold, as it retains more warmth than if it were any other hue; and, with the exception of the species under consideration, the members of this genus are inhabitants of the most inhospitable portions of our globe—delighting in the fierce blast, and making their abodes amid the deep snows of the loftiest mountains. Now, for the Scotch Grouse this change of plumage was unnecessary, as it rarely ascends higher than two thousand feet above the sea, but lives mostly in a comparatively mild climate, where its ordinary garb was sufficient protection, and the extra precaution of a white mantle unneeded. It seems that the mild climate is the most satisfactory reason which can be given for its not changing the color of its plumage with the season; since otherwise, being a Ptarmigan, it ought to change as regularly as its relative, the Lagopus Mutus, which abides upon the slopes and summits of the high mountains that look down upon the heath-clad plains.

It is singular that this bird should be so nearly allied to the Lagopus Albus, and yet present sufficient characters to entitle it to a specific distinctness. Singular, because, in examples of the Willow Ptarmigan from Lapland, Norway, Sweden, and throughout the northern portions of the American continent, I can as yet discover no differences between them worthy of constituting a separate species; although leagues of ocean and of land may have divided their various haunts during life; yet this species is only to be found in the British Islands. The one encircles nearly the entire globe; the other is confined within narrow bounds.

The fact of the present species not turning to white in winter, does not invalidate its claim to be considered as generically a true Lagopus; and to sustain this opinion, there is ample evidence of the same fact existing among the quadrupeds, when some species of the same genus turn white in winter in some latitudes, while others, inhabiting different climates, do not; yet no one, for this cause, would desire to arrange them under separate genera.

In looking at this subject, it must also be taken into consideration that the egg of the present species is strictly of the Ptarmigan style, and not of the Grouse.

It would thus appear, that, with our present means of judging, the Scotch Grouse should be held as a true Lagopus, but specifically distinct from the L. Albus.

The male has the head, neck, breast, and sides bright chestnut, irregularly crossed with fine black lines. The middle of the breast, and abdomen, very dark brown, sometimes black, with many of the feathers tipped with white. Under tail coverts chestnut with a terminal white bar. The upper parts are less bright than the lower, the feathers transversely barred with black, and frequently having patches of black, with fine bars of yellowish white. The primaries and secondaries are chocolate brown, the outer webs of the latter minutely mottled with reddish brown. The upper tail coverts are like the back, sometimes having white tips. The tail has the two centre feathers chestnut, barred with black, the next two more slightly barred, and the remainder of a dark chocolate brown. The feathers of the tarsi are brown, but much lighter on the toes. Bill black, with a white spot at the base of the lower mandible.

The female is much lighter than the male, the general color being of a yellowish brown, with the transverse markings and patches much more distinct. The breast is reddish brown barred with black. The white spots at the base of the bill are not as clearly defined as in the male; the primaries are chocolate brown, as are the secondaries, the latter more broadly mottled. The feathers of the tarsi and toes are pale gray. The bill black.

The young are covered with a yellowish down, marked on the back and sides with dark brown, and having the top of the head chestnut, with a spot before and behind the eye of a dark brown. Bill brownish black; the claws pale brown.

The plate represents the male, female, and young of the natural size.

ROYAL GROUSE.
LAGOPUS PERSICUS

LAGOPUS PERSICUS.   G. R. GRAY.

KUNALEE GROUSE.

LAGOPUS PERSICUS.   Gray, Gen. of B., vol. iii., pl.—Id., Cat. B. Brit. Mus., Pt. III., p. 48 (1844).

This bird, figured by Gray in the work above quoted, is said to be a native of Persia.

It is hazardous to announce a specimen of Ptarmigan as belonging to an undescribed species, unless ample opportunities have been afforded to compare it with others from the same locality, and which may also show like variations from well-known forms, since the members of even the same species in this genus present differences both in color of plumage and in the measurements of their parts, greater than may be found in perhaps any other class of birds.

If one may judge by the lifelike portrait in the accompanying plate—the result of Mr. Wolf's unrivalled skill—this bird bears a strong resemblance to the *Lagopus Scoticus*. It is indeed of a lighter color than the typical examples of that species, yet we know that the Scotch Grouse vary very much, in different localities, in their plumage, and it would not be deemed an unusual occurrence to find one of as light a hue as that in the illustration.

As the Ptarmigan are natives of northern climes, it may be considered as somewhat strange that one should be discovered in Persia; and therefore it would seem desirable, before admitting this bird to rank as an undoubted species, that more specimens should be procured from the same country, and that they also should present a like peculiar style of coloring in their plumage.

I am without any information in regard to the economy or habits of the Kunalee Grouse, but presume that it would in both resemble the *Lagopus Scoticus*.

Mr. Gray gives Kaipariah Persia as the locality whence this specimen came.

The drawing of this bird, which gives us so much better an idea of it than the most minute description could, was made from the specimen in the British Museum, and which is the only one, I believe, that has ever been obtained.

LAGOPUS MUTUS.

TETRAO LAGOPUS. Linn., Syst., vol. i., p. 274, sp. 4.—Gmel., Syst., vol. i., p. 749.—Temm., Man. d'Ornith., vol. ii., p. 468.—Gravos, Br. Ornith., vol. ii.—Naum., Vog. Donts. (1833), vol. vi., p. 401.—Rich., Supp. Parry, 2d Voy., p. 350.—Sab., Appen. Parry, 1st Voy., p. cxcvii.—Jenyn., Man. B. Vert. Anim., p. 170.—Schleg., Rev. Crit. des Ois. d'Eur., p. 76.—Lath., Ind. Ornith., vol. ii., p. 639, sp. 9.—Fab., Faun. Groenland., p. 114, sp. 80.—Meyer & Wolf. Deut. Vogel, Vogel, p. 296, vol. i., pl.—Hahn, Deutsch., Vog., p.—Brehm., Ornith., Beitr., H. 3, p. 252.

PTARMIGAN. Lewin's Br. Birds, vols. v. and vi., pl. 135.—Lath., Syn., vol. iv., p. 741, sp. 10.—Montag., Ornith. Dict., vol. ii.—Thomp., Nat. Hist. Irel., vol. ii., p. 45.—Morris, Hist. of Br. Birds, vol. iii., p. 351, pl. 172.—Penn., Br. Zool., 1812, vol. i., p. 350, pl. 57.—Yarr., Brit. Birds, vol. ii., p. 322.

TETRAO RUPESTRIS. Gould. Birds of Eur., pl. Female.

LAGOPUS VULGARIS. Vieill., Nouv. Dict. d'Hist. Nat., vol. xvii., p. 199.—Flem., Brit. Anim., p. 43.—Eyton, Cat. Br. Birds, p. 30.

WHITE GROUSE. Penn., Arctic Zool., vol. i., p. 300.—Bewick, Br. Birds, vol. i., p. 353.

LA GELINOTTE BLANCHE. Buff., Pl. Enl., t. 129, 494.—Briss., vol. i., p. 216.

LE LAGOPEDE ORDINAIRE. Cuv., Reg. Anim., vol. i., p. 482, (1829).

LE LAGOPEDE. Buff., Nat. Hist. Ois., vol. ii., p. 301.

TETRAS PTARMIGAN. Temm., Pig. et Gallin, vol. iii., p. 185, t. Anat., 10 f. 1, 2 et 3.

LAGOPUS CINEREUS. McGill., Brit. Birds, vol. i., p. 187.

LAGOPUS ALPINUS. Nils., Skand. Faun., vol. ii., p. 98.—Keys et Blas. Wirbeth. Eur., p. 63.

TETRAO ALPINUS. Nils., Orn. Suec., vol. i., p. 311.—Gloger., Voy. Eur., p. 553.

TETRAO MONTANUS. Brehm, Voy. Deutsch, vol. ii., p. 448.

LAGOPUS MUTUS. Leach, Cat. of Mam. and Birds, Brit. Mus., p. 27.—Gray, Gen. of Birds, vol. iii.—Ib., Cat. Birds Brit. Mus., p. 48, (1844).—Selby, Brit. Ornith., pls. lix. and lx., figs. 1 and 2.—Steph., Gen. Zool., vol. xi., p. 287, pl. 21.—Bonp., Rev. de l'Ornith. Eur., p. 173.—Gould, Birds of Great Britain, Pt. vi.

The common Ptarmigan is a native of the northern portions of the Old World; and, on account of the many changes of plumage to which it is subjected, has been called by many different names, founded seemingly upon the various localities from which the specimens came, rather than upon the sufficient reasons usually given for constituting separate specific distinctions.

So many are the changes of dress which these birds undergo, and so rapidly do these follow each other, and so astonishingly do individuals vary from each other, that it is utterly impossible to give a description of the Lagopus Mutus, at any season of the year, save winter, which would answer for the entire species. Changing and ever changing, in a continual state of moult, the feathers are no sooner perfected than they are obliged to give way to others of different hue; and thus, when this moulting, from age or other causes, does not commence and terminate in all individuals at the same time, it may readily be imagined how difficult is the task to define the species, when one is restricted to the color and markings of this mutable, evanescent plumage.

The Ptarmigan makes its abode upon the loftiest mountains, descending in summer into the valleys only for a short period to breed. MacGillivray says: "The nest is a slight hollow, scantily strewn with a few twigs and stalks, or blades of grass. The eggs are of a regular oval form, about an inch and seven-twelfths in length, an inch and from one to two-twelfths across, of a white, yellowish-white, or reddish color, blotched and spotted with dark brown, the markings larger than those of the Red Grouse. The young run about immediately after leaving the shell, and from the commencement are so nimble and expert at concealing themselves, that a person who has accidentally fallen in with a flock very seldom succeeds in capturing one. On the summit of one of the Harris Mountains, I once happened to stroll into the midst of a covey of very young Ptarmigan, which instantly scattered and in a few seconds disappeared among the stones, while the mother ran about within a few yards of me, manifesting the most intense anxiety, and pretending to be unable to fly. She succeeded so well in drawing my attention to herself, that when I at length began to search for the young, not one of them could be found, although the place was so bare that one might have supposed it impossible for them to escape detection."

In regard to the changes of plumage which this species undergoes, I quote the words of Mr. Wheelwright, than whom no one has had better opportunities for observing these birds, in Sweden and Lapland. He says: "When we first arrived on the fells (April 10), some of the Ptarmigan were still in pure white winter dress; others were just beginning to assume the summer plumage, and here and there a summer feather was shooting out on the head and neck. In about a month's time many of the summer feathers had appeared in different parts of the body of both males and females, and about May 22d the ovaries of many of the females were in a very forward state, but the change in plumage seemed to go on slowly. On June 5th we took our first nest, with ten eggs, and the old female (which I shot as she rose), showed nearly as much of the winter as of the summer plumage. By June 10th the males were, however, grayish-black on the head, back, and chest, the belly and under part pure white; the black color darkest on the breast. The change from the winter to the summer dress is a true moult, and not a change of color in the feathers. It is most difficult to say what is the real summer dress of the Ptarmigan, for they appear to be in a continual state of change or moult during the whole summer, and bear no one dress for any length of time; and so irregular is the moult or change, that you scarcely ever see two exactly alike or in the same state of forwardness, for in the same day in July you may kill some in the early summer dress, and others with many blue autumn feathers. Up to July 9th I observed that all the old males which I killed were dark brownish-black on the back, speckled with lighter brown, especially on the head, breast, and sides; belly, pure white, but the dark breast is much more conspicuous in some than in others. By July 20th the whole body-color had become lighter, and by the end of July was evidently changing to blue-gray, but still speckled with brown, especially on the head. By the 6th of August the majority of the males had assumed a totally different dress: head still speckled with yellowish-brown; back, bluish-gray, watered with black and white; belly, pure white; and this was the plumage of the males on August 18th, when I killed my last. This blue-watered dress appears by degrees to become fainter, in fact more gray-blue, until the end of September; but the white winter feathers keep gradually showing themselves under the blue autumnal dress. I observed in two specimens shot early in October the year before, that one was half blue and half white, i. e., that the body appeared to be covered with the blue autumn dress, the other half with the white winter plumage, some of which, if not all, were perfectly now feathers, for I observed blood-shafts to many of them; in the other specimen, very few of the blue autumn feathers remained. From what I could hear, for I did not stop up long enough to judge for myself, I should say that in many, perhaps most, the pure white winter dress is complete by the third week in October.

"Much as the males vary in plumage, the females appear to vary still more, and only to have a standing dress for about three weeks in June, just when they are laying, and this early summer dress may be described thus: body, blackish-brown, every feather broadly edged with yellow, brown, and white, giving the bird a very light yellow-brown appearance; breast, much lighter; belly, never pure white, as in the male, but, as well as the sides and breast, covered with black zig-zag lines on a rusty yellow and white ground, the white color most apparent on the belly. By the second week in June this dress was complete in most, although the birds vary much in shading, scarcely two being exactly alike, when it all at once became much darker. In fact, we may describe the summer dress of the female Ptarmigan thus: throughout the whole of May the ground plumage was white, here and there speckled with mottled rusty yellow and black feathers, which, as in the males, appear first on the head and neck, then on the back. By the third week in May the body is thickly speckled with these mottled feathers (some intermingled with the white, others shooting out from the skin under them), so we are not at all surprised that early in June a sudden change takes place, and all at once the bird assumes its early or first summer dress, as above described. About the end of July we see some small blue feathers shooting out among the rusty brown ones, and this appears to be a true month, and not a change in color of the feathers. The bird now assumes a beautiful dress, far more handsome than the male—brown-red, variegated with blue-gray, which often on the back appears in patches. But the females vary so much in color that a minute description of one would not apply to another. I fancy that both male and female retain this blue dress longer than any other. It gradually becomes lighter as the season advances, till at length the old female is quite blue (but still always with some rusty mottled yellow feathers at the sides), and about the middle of October the blue dress gives place to the pure white plumage of winter.

"The plumage of the young in the downy state is rusty yellow, with longitudinal markings and minute spots of black; the first dress after that is black, mottled with rusty yellow and white above; underneath, pale rusty brown with blackish wavy lines; wings, grayish-brown. Early in August the body plumage becomes grayish-blue, finely streaked with black, and the pinions white instead of brown. This gray plumage gradually becomes lighter, as in the old birds, till, like them, they assume their winter livery, and by the first of November there is no perceptible difference between old and young birds.

"It appears, then, that the Swedish Ptarmigan has three distinct dresses in the course of the year, and so many intermediate changes that they appear to have a different dress for every summer month.

"The Ptarmigan may truly be said to be a child of snow, for you never meet with them off the real fells, although I have occasionally flushed them from the fell-sides, just where the willow bushes end. Their real home is the higher fell-tract, and in the middle of summer on their very highest snow-clad summits. In the spring they come down to the lower fells to breed, but you never find them there in the end of summer. The pairing season hence appears to begin early in May, and lasted a fortnight or three weeks, and during this time the hoarse laughing love-call of the old male might be heard at very earliest dawn on any of the fell-tops. This is soon answered by the finer 'i–i–aek; i–k—aek,' of the female, and the love-chase commences. This is the time when many are shot off, for they are now too engrossed with each other to heed the shooter, who lies behind a stone on the pairing ground, and picks them off as he pleases.

"Both the Ptarmigan and the Willow Grouse are strictly monogamous. Some naturalists appear to have an idea that both, when pairing, have a kind of 'lek,' or play, like the capercailzie and black cock, both of which birds are polygamous. I can only say I never saw anything of the kind. The Ptarmigan certainly have their favorite pairing grounds on the fells, and here the birds assemble at daylight in the early spring, in small flocks, but widely scattered all over the place. The old males utter their peculiar love-call, which is answered by the female, and they draw together; but although there are several males in the neighborhood, each one seems to have his particular stand and his own favorite female, and if by chance another male intrudes on his ground, he drives it off.

"Early in June the female commences to lay, forming an artless nest on the bare stones in the heather, or under a small bush; always, as far as I could see, above the very top edge of the willow region, but never on the snow-fells. As long as the female continues to sit, the old male watches in the vicinity of the nest, like the Willow Grouse; but as soon as the young are hatched off he leaves them to the care of the mother, and joining a lot more 'bachelor friends,' they seek the tops of the highest fells (leaving the female and young brood lower down in the fell valleys). Early in August the young will be strong flyers; the old female then takes them higher upon the fells; they are joined again by the old male, and the whole family keep together till the autumn snow falls, when several families pack, and large flocks are met with in the lower fell-tracts during the whole winter.

"In the summer the food of the Ptarmigan seems to consist entirely of leaves, flowers, and fruit of the fell-shrubs. The young live much on insects, and in the winter the frozen fruit of the crowberry and cranberry afford them ample supply of food, and there are always bare places, even on the highest fells, from which the wind has blown the snow."

Some authors have stated that they have met with this species in Arctic America, and that the specimens obtained differed in no way from the Ptarmigan of Scotland. In the collections sent to the Smithsonian Institution, obtained in the northern portions of the American continent, I have never seen any specimens of the Lagopus Mutus, the only species of Ptarmigan represented being the Lagopus Albus, and Lagopus Rupestris; consequently, I am strongly inclined to the belief that the common Ptarmigan of Europe is not found in the New World, but is represented by the Rock Grouse.

I have found it necessary to give two plates of this species, in order to show the change of plumage the bird undergoes. The first represents adults and young in summer; the male in the back-ground already beginning to assume the fall style of coloring. I have observed this change to occur in individuals quite early in the summer. The other plate represents the birds in their winter dress. All the figures are of the natural size.

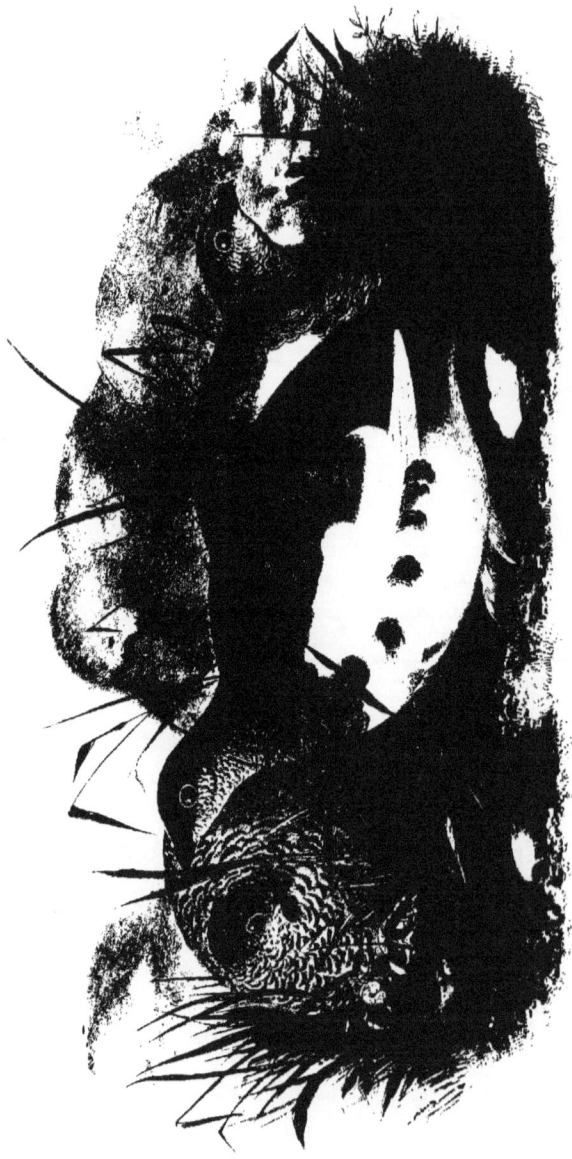

LAGOPUS RUPESTRIS.

TETRAO RUPESTRIS. Gmel., Syst. Nat., vol. i, 1788, p. 751.—Lath., Ind. Ornith., vol. ii., 1790, p. 640.—Sab., Supp. Parry, 1st Voy., pl. cxiv.—Rich., Appen. Parry, 2d Voy., p. 348.—Aud., Ornith. Biog., vol. iv., 1838, p. 483, pl. 368.—Swain, Faun. Bor. Amer., vol. ii., 1831, p. 354, pl. xiv.

TETRAO LAGOPUS. Temm., Greenl. Birds, No. 4. p. 468.

ATTAGEN RUPESTRIS. Reich., Av. Syst. Nat., 1851. pl. xxiv

ROCK GROUSE. Penn., Arct Zool., vol. i., p. 364, and vol. ii., p. 312.

LAGOPUS RUPESTRIS. Leach, Zool. Misc., vol. ii., p. 290.—Aud., Syn., p. 208.—Id., B. of Amer., vol. v., 1842, p. 122, pl. 301.—Gray, Gen. of B., vol. iii.—Baird, U. S. P. R. R. Exp. and Surv., vol. ix., p. 635,—Bon., Geog. and Comp. List Birds, p. 44, No. 290.

LAGOPUS AMERICANUS. Aud., Syn., 1839, p. 207.—Id., Birds of Amer., vol. v., 1842, p. 119, pl. 300.—Baird, Birds of N. Amer.

LAGOPUS ISLANDORUM. Fab., Prod. der Island. Orn., p. 6.—Gray, Gen. of Birds.—Id., Cat. B. Brit. Mus., Pt. III, p. 47, 1844.

TETRAO LAGOPUS ISLANDICUS. Schleg., Rev. Crit. des Ois. d'Eur., p. 76.

TETRAO ISLANDICUS. Brehm., Eur. vogs., vol. ii., p. 448.

LAGOPUS REINHARDTII. Brehm.

LAGOPUS GRAENLANDICUS. Brehm. Vogelfang, p. 264, note.

This species appears to be found only in Iceland, Greenland, and the northern portions of the American continent, and is not, so far as my investigations show, an inhabitant of the Old World.

It is closely allied to the common Ptarmigan; but I have never seen, in any of the specimens of Lagopus Mutus, that I have had the opportunity of examining, the peculiar markings and coloration which characterize the present bird.

Specimens of the Rock Ptarmigan lately received from Arctic America through the collectors of the Smithsonian Institution, in no wise differ from many before me from Greenland and Iceland: while from their larger and differently shaped bills, and the yellowish-brown hue of their plumage, they all would seem to be entitled to a specific distinction from the Lagopus Mutus.

I have therefore considered the Lagopus Islandicus, Lagopus Reinhardtii, Lagopus Graenlandicus, and Lagopus Americanus as synonyms, as the term Rupestris takes precedence of them all.

The Lagopus Americanus of Audubon may, with some degree of certainty, be considered as the present species; for although he gives no distinctive characters to separate it from either the Lagopus Albus or Lagopus Mutus; yet as he states its total length to be only fourteen inches, and says that his specimen was brought from North America, it may reasonably be supposed to be the Lagopus Rupestris in change.

The Rock Ptarmigan undergo similar changes in their plumage, as is customary with the common Ptarmigan. In winter, with the exception of the tail, which is always black, the entire plumage is white, the males being distinguishable from the other sex by a black mark through the eye. Hearne says of this species that "they never frequent the woods or willows, but brave the severest colds on the open plains. They always feed on the buds and tops of the dwarf birch, and after this repast generally sit on the high ridges of snow, with their heads to windward. They are never caught in nets like the Willow Partridge, and being so much inferior in size, their flesh is by no means so good, being black, hard, and bitter. They are in general like the Wood Partridge, either exceeding wild or very tame; and when in the latter humor I have known one man kill one hundred and twenty in a few hours; for as they usually keep in large flocks, the sportsman can frequently kill six or eight at a shot.

"Like the Willow Partridge, these birds change their plumage in summer to a beautiful speckled brown; and at that season are so hardy that unless shot in the head or vitals, they will fly away with the greatest quantity of shot of any bird I know. They discover great fondness for their young, for during the time of incubation they will frequently suffer themselves to be taken by hand and off their eggs."

The plate represents two males and a female of the natural size.

LAGOPUS HYPERBOREUS.

SPITZBERGEN PTARMIGAN.

LAGOPUS HYPERBOREUS. Journ. für Ornith. (1863), p. 371.
LAGOPUS ALPINUS—var HYPERBORREA. Gaimard, Voy. en Scand., 18m., Livrais.
TETRAO LAGOPUS. J. C. Ross, in Parry's Attempt to reach the North Pole, 1827, S. 193.
FISVOGEL MARTEUS. Spitz. Reiseb. S., 53.
LAGOPUS HEMILEUCURUS. Gould, Proc. Zool. Soc., 1866, p. 354.

This bird, although closely allied to the species I have designated as Lagopus Rupestris, yet on account of its great size would seem to be entitled to a specific distinctness. It was first obtained by Professor Sundvall (who accompanied the expedition of Gaimard), at Bellsund, about 77° 40' latitude, on the first of August in 1838. It was a male, and the only one obtained at that time.

The next specimen, a female, was procured by Mr. Evans, and described by Mr. Gould as Lagopus Hemileucurus, on account of the basal portion of the tail-feathers being white. The following note was furnished to the above eminent ornithologist by Mr. Evans, and comprises all that is known regarding this example:

"The skin sent is the only one I have from Spitzbergen, although I shot many. The birds were so plentiful that thinking I could always procure examples, I neglected to preserve any at the time, and was obliged to come away at last with only this one.

"The hen-birds had all assumed their summer plumage, but the males had not changed a feather, though the old ones, which had become very ragged and dirty, would almost fall off on being touched. I started one hen from her nest, or rather from the little dry hollow where she had collected a few stems of grass, and found two eggs; these were all we met with; the nest was placed in the high fields, where, in the dry parts, scarcely any vegetation is to be seen, while the swampy portions, where the snow had melted, were covered with coarse grass and the dwarf willow, which is the only thing approaching to a shrub on these barren, treeless islands. The specimen sent was shot on the 27th of June, on the south shore of Ice Sound, in about 78° north latitude.

"The neighboring country consisted of a belt of swampy ground covered with rank grass, with high, rugged, and barren mountains rising behind, covered with snow, except on their sharp ridges and steep sides; these mountains, which are interspersed with vast snow-clad plains, stretch away for miles inland, and rise with beautiful cones in the distance; here and there, in a few sheltered spots, a scanty supply of small flowers is to be found, mostly belonging to the following families: Draba, Ranunculus, Saxifraga, etc. The dark-gray rocks were covered with lichens in great variety, but of a gloomy and sombre hue, in strict keeping with the wildness of the scene; here, too, the reindeer-moss grew in great abundance. I may remark that the Ptarmigan were so tame, that we could easily have knocked them down with a long stick—doubtless from being so unaccustomed to the intrusion of human visitors."

I now give an extract from a letter written by Professor Sundvall to Professor Baird, replying to some interrogations made by the latter, at my request, regarding the specimen of this bird contained in the museum at Stockholm:

" * * * * * As for the Spitzbergen species, I think there is no doubt that there is only one. The Lagopus Hemileucurus of Gould (1858) must be the same as my Lagopus (Alpinus var) Hyperborea in Gaimard's Voyage (published before 1847, or, as I remember, in 1845). The only difference in the description is, that Gould says the rectrices are half part white, but in our specimens they are scarcely one-third of their length white; which difference may arise from age or sex, as Gould's is a female, and ours are all males. Our museum contains three specimens. The first, a male in summer plumage, shot 1st of August, 1838, at Bellsund, about 77° 40' latitude, and prepared by myself. It was the only one seen during the expedition, although we remained ten days at Bellsund. Of this specimen there is an excellent figure in Gaimard's Atlas, drawn and colored here in Stockholm by one Wilhelm von Wright. The specimen itself is in a bad state, as the bird was moulting, and it has only one of the rectrices (the extreme left), which is 130 millimetres in length, but broken; and it has lost nearly all the remains of the white apical margin. The base is only about 40 millimetres on the external web, but blackish on the whole inner. The scapus is 18 or 20 millimetres long, whitish. Length of wing, 220 millimetres. The total length is now, as the bird stands with curved neck, only about 370 millimetres. In the figure, the length is 175 millimetres, but if the neck was more stretched, as it would be were the dead bird laid on a table, the total length would be more than 400 millimetres. Tarsus, 37 millimetres; middle toe, 25, and with claw, 41 millimetres. The wing is white, but the last remiges, tertials, and a number of the greater rectrices towards the back, are new summer feathers, as shown in the plate. The scapi of the remiges are white, with a fuscous or blackish stripe *along the middle*, but terminating long before the end of the feather is reached. The rest of the coloring does not differ from the males from Greenland and Iceland with which I have been enabled to compare it. The only fault in the figure I have mentioned, is, that the lores are not dark enough, which is caused by the fact that the greater portion of the black feathers are out, and the new ones had not been perfected.

" Within a few years we have obtained two other specimens: one, a male in *pure winter dress*; white, with broad, perfectly black lores; shot on north coast of Spitzbergen, 1st of June, 1861, and brought home with Thorell's expedition. It is in a bad state, but the tail is complete. Besides the two white rectrices there are seven black to each side: 150 millimetres long, and the base, for 50 millimetres, white, with the shaft as in the first specimen, but whitish for about 20 millimetres nearer the end of the feather. The remiges, like the former, have a dark middle stripe along the shaft. Dimensions a little greater. Total length, moderately stretched, 460 millimetres; wing, 228; tail, 150. It is thus larger than Gould's female.

"Our third specimen, a male, was beginning to moult,—shot the 7th of July, 1864, at Icefjord, and brought home by Malmgren. It is also in poor preservation, and is white; but the head, neck, regis scapularis, have very many new summer feathers. Tail and remiges exactly like those of the two other specimens, but dimensions a little greater. Length of wing, 235 millimetres; tail, about 150; bill and nails, blackish, as in the two former. The white on the base of the tail is concealed by the surrounding feathers in all three specimens.

"All are in a bad state, as the two latter expeditions could not remain a long time in each place, and the skins dry very slowly in that climate. My own specimen would have been better if it had not been moulting, with most of its feathers *blood feathers*. This bird seems to be scarce at Spitzbergen, in all three instances. In ours there were no more obtained than the one brought home, and only at Icefjord did Malmgren see two more. Mr. Gould's ornithologist says he found them very plentiful; but he probably only met one somewhat large family, which he has stoutly destroyed. A great number of travellers, who shoot only to kill, or perhaps to eat, contribute very much to perfect the work of the ice-foxes at Spitzbergen, and of the common foxes in other places.

" In the specimen brought home by me there were only plants in the œsophagus mucous, as leaves and flowers of Saxifrage, etc. Malmgren states (in the Rev. of the Acad. Sc., of Stockholm, 1864, p. 379), that he once heard a sound uttered by the male like *arrr* or *errr*, in a coarse voice, resembling somewhat the croaking of a frog. Fabricius also remarks this in the Greenland species.

"On comparing these birds with the males from Greenland and Iceland, these last are found to be much smaller (wing 190 and 193 millimetres), and the base of the rectrices much less white, which color does not extend farther on the shaft than on the web; also, the shafts of the remiges are black for their whole breadth. As these differences seem to be constant, they are sufficient to render the Spitzbergen bird always recognizable from the other two, and thus entitle it to be considered a distinct form, if we may not even believe it to be of different origin.

"I have a female from Greenland, and in this the white basil part of the outer rectrices has really a little difference in form from the males. It is larger on the outer side. From the European Lagopus Mutus they all differ, evidently, in the males more, the females from Greenland less, in color, but they come very close to it in the form of the bill, black lores, etc."

As it seems pretty evident that the extent of the white on the tail varies considerably in different specimens—a fact which I have noticed in a large number of examples of Lagopus Albus—the claims of this bird for specific distinction rest upon its large size, which, at the best, is a very questionable sufficiency; and it would seem most likely to be the Lagopus Rupestris: but without any number of examples to enable me to form my opinion, I have deemed it best to give a figure of the female sent to Mr. Gould, and to hope that some no very distant day will afford the material for rightly determining what is now so doubtful a point.

WHITE-TAILED PTARMIGAN.

LAGOPUS LEUCURUS.

LAGOPUS LEUCURUS. Swain.

WHITE-TAILED PTARMIGAN.

TETRAO (LAGOPUS) LEUCURUS. Swain and Rich., Faun. Dor. Amer., vol. ii., 1831, p. 356, pl. lxiii.—Nutt., Man. Ornith., vol. ii., 1834, p. 612.—Ib. vol. i, 2d edit., 1840, p. 620.—Doug., Trans. Linn. Soc., vol. xvi., p. 146.

TETRAO LEUCURUS. Aud., Orn. Biog., vol. v., 1839, p. 200, pl. 418.

LAGOPUS LEUCURUS. Aud., Syn., 1839.—Ib. Birds of Amer., vol. v., 1842, p. 125, pl. 302.—Gray, Gen. of Birds, vol. iii.—Baird, U. S. P. R. R. Exp. and Surv., vol. iv., p. 637.—Bon., Goog. and Comp. List Birds, p. 44, No. 291.—Elliot, Proceed. Acad. Nat. Sciences (1864), p.

It is with much gratification that I am enabled (through the kindness of Professor Baird, who, with his accustomed liberality, has placed in my hands the large collection of American Grouse and Ptarmigan belonging to the Smithsonian, to assist me in my investigations for this work,) to give a representation of this species in its full summer plumage. Heretofore it has only been known to us by descriptions, or by one or two mutilated specimens in the winter dress, and only lately have examples been received, as represented in the plate. It is an inhabitant of the lofty peaks of the Rocky Mountains, and of the snowy heights that look down upon the Columbia River. Like all the true Ptarmigan, this species turns white in winter, and is readily distinguished from all its relatives by having the tail always of a pure unmixed white.

This species was first obtained by Mr. Douglas, but he failed to bring his specimens home with him. He says in his paper: "But in the first place I may be permitted to mention a new species, nearly allied to *T. Lagopus*, but much smaller, with a white tail, and when in winter dress, snow white, without the least particle of black. This is an inhabitant of the Rocky Mountains, and the snowy peaks of Northwest America. During my journey across the dividing ridge in April, 1827, I killed several, which, from the extreme difficulties to be surmounted at that early season of the year, I was reluctantly obliged to leave behind me. This loss I do not now regret, as Dr. Richardson was fortunate enough to secure the species, an accurate description of which will be shortly given by him in his forthcoming 'Fauna of British North America.'" Richardson, in the work above mentioned, says: "Of this species I have only five specimens, four procured by Mr. Drummond on the Rocky Mountains, in the fifty-fourth parallel, and one by Mr. Macpherson on the same chain, nine degrees of latitude farther north. * * * * The sexes of my specimens were not noted, but none of them have the black eye stripe; and Mr. Drummond, who killed great numbers, is confident that that mark does not exist in either sex."

The habits of the White-tailed Ptarmigan are said, by the authors quoted above, to resemble those of the Ptarmigan (*Lagopus Mutus*). Preferring the temperature of eternal snow, they descend to the lower portions of the mountains only for the purpose of incubation, and return again to their loved mountain tops as soon as that duty is accomplished. These birds are admirably adapted by nature to withstand the most intense cold, being so densely and completely covered with feathers as to leave only the bill and ends of the nails exposed to the piercing blasts which sweep over their snow-clad homes. The change of plumage also is an additional protection given to them by the all-wise Creator, for clad in their winter dress of pure white, they are so assimilated to the snow around, as to render them invisible, even to the searching eye of the hungry hawk; and in summer, by approaching to the hues of the lichen and moss, they are almost impossible to be distinguished from a clod or turf as they nestle closely to the ground. They, like all of the Lagopidæ, do not commence to change from the summer to the winter dress, and *vice versa*, at the same time, in all individuals; and thus it is difficult to find two examples exactly alike. They also vary in the color of their plumage, some being much darker than others, with broader bars upon the feathers; sometimes even having large blotches of black upon the back. It is this difference in the color of the plumage of individuals which has caused so much confusion in the classification of this family, and specimens have been described as new species, which eventually would prove to be but varieties. This, however, has not been the case with the present species, as its white tail would at all times clearly separate it from every acknowledged species of Ptarmigan.

If we would see them in their haunts, we must climb to the heights whereon they dwell;—perhaps no easy task,—but the student of nature must incur fatigue, and overcome many obstacles, before he can acquire the knowledge which she seems so often to love to conceal. Up the rugged sides of a lofty mountain, whose summit is clothed in perpetual snow, and rich in prismatic hues, our path lies; and as, leaving the plain, we gradually ascend, the landscape, unfolding itself beneath us, with many checkered colors, is lost in the far horizon. The streams, in their wandering course, glisten like silver threads upon the rich carpet through which they flow, and the entire view lies bathed in the mellow light of the sun's rays. But we still go on, and the breeze coming from the snow-clad fields above, chills like the breath of winter; while the hills that seemed so lofty when we commenced to ascend, appear now like slight undulations of the soil, as little perceptible as the heaving breast of the slumbering ocean on a calm midsummer's day. All around is still; no sound breaks the silence save our own footfalls as we straggle on, or perhaps the scream of the startled hawk, as he wheels in circles over us. As yet we have not seen a Ptarmigan, and the sparse vegetation around does not seem capable of supporting bird-life to any extent; but suddenly, springing from almost beneath our feet, one rises from the moss and tufts of grass where it had lain concealed, and flying only a short distance, alights upon some projecting rock, where, after having watched us for a few seconds, standing perfectly motionless, it commences to dress its feathers, apparently taking no further notice of our movements. Before we proceed, let us cast our eyes around, and we may find the companions of the one before us, for the Ptarmigan loves the society of its own species, and is rarely found alone. At first nothing but the stones and grass meet our gaze, but yonder is a clump of grayish hue, which, as we draw nearer, takes a more definite shape, and from the midst of its compactness, twinkle a pair of bright eyes all alert to our movements. Drawing still closer, it stirs, and rising on sounding pinions, discovers the living, vigorous bird; which, with easy flight, joins its mate before us. And now, our eyes more accustomed to distinguish their forms, we see them on every side nestled closely to the ground; and in order that they may recover their confidence, let us return a short distance, and seat ourselves. Soon a faint chirp is heard, and several little heads are raised, and one individual bolder than the rest runs a few steps, then stops and looks around,—an insect flying over attracts the eye of one, and he springs to catch it, and is joined in the pursuit by several more. Thus, one by one, they return to their usual occupations, some seeking seeds, others dusting themselves in the way, all fear of our presence having been removed; and thus gratified with beholding them pursue the daily callings of their peaceful natures, we will leave them, a happy, contented little society, and turn on our downward path.

In winter the present bird is perfectly white, never having in either sex the black mark through the eye observable in the males of perhaps its nearest ally, the *Lagopus Mutus*, and differs from that species also in its tail being always white, instead of black tipped with white. In summer the head and back part of the neck is crossed with fine lines of black and yellow, the feathers on top of the head tipped with white. Entire upper parts golden gray, spotted with iron gray, and confusedly mottled with black. Feathers on fore part of breast darker than the back, the black more conspicuous, and a broad white spot in the centre on both sides of the shaft, this, however, not extending to the tip. Wings, lower part of breast, and tail, white at all seasons. In winter the plumage is pure white.

The plate represents the two sexes the natural size.

1–2. Tetrao Urogallus.–3–4. Lyrurus Tetrix.–5–7. Bonasa umbellus.–8–10. Dendragapus Richardsonii.–11–12. Centrocercus Urophasianus.–13–17. Canace canadensis.–18. Canace Franklinii.–19–21. Pedioecetes Columbianus.–22–24. Pediocaetes Phasianellus.–26–30. Cupidonia Cupida.

PLATE I.

The eggs of the various species of this family differ considerably from each other, both in color and markings, and this variation is observable also, even among those in the same nest. Among the Lagopidæ there does not seem to be any typical style of marking to designate the species; as I have never been able to find any two eggs, even when from the same nest, exactly alike; therefore it has seemed desirable to figure several examples of one species so as to exhibit as many of the most striking varieties as possible. The numbers which are given below in large type are those belonging to the catalogue of the Smithsonian Institution; where the types of these plates are preserved for future reference.

TETRAO UROGALLUS. Nos. 1, 2 (**5330, 1378**) were presented by Alfred Newton, Esq., to the Smithsonian Institution, and were obtained from Finland.

LYRURUS TETRIX. Nos. 3, 4 (**1455, 3547**), also presented by Mr. Newton, were brought from Germany.

DENDRAGAPUS OBSCURUS. Nos. 5, 6, 7 (**2583, A. 2583, B. 2583, C.**) were procured by Dr. Kennerly in Washington Territory.

DENDRAGAPUS RICHARDSONII. Nos. 8, 9, 10 (**3888, A. 3888, B. 3888, C.**) were collected by Captain Reynolds in the Wind River Mountains, lying to the northwest of Fort Laramie.

CANACE CANADENSIS. Nos. 13, 14, 15, 16, 17 (**3091, 7616, 7613, 7612, 7615**) were brought by Mr. L. Clarke from Fort Rae, Great Slave Lake.

CANACE FRANKLINII. No. 18 (**4694**) was obtained by Mr. G. Gibbs in Washington Territory. Only three specimens of the egg of this species have been discovered.

PEDIŒCETES COLUMBIANUS. Nos. 19, 20, 21 (**5239, 3810, 5238**.) The first two came from the neighborhood of Fort Crook, California, and the last from Missouri River.

PEDIŒCETES PHASIANELLUS. Nos. 22, 23, 24, 25 (**7618, 7620, 7619, 7621.**) The first was procured at Fort Resolution, south side of Great Slave Lake; the remainder at Fort Rae, on the north side of the lake.

CUPIDONIA CUPIDO. Nos. 26, 27, 28, 29, 30 (**7027, 7023, 5432, 1880. 4011.**) All of these were obtained in Illinois, excepting the last, which was brought from Arkansas.

PLATE II.

LAGOPUS ALBUS. Nos. 1, 2, 3, 4, 5 (7535, 7633, 7640, 7630, 7694) were obtained at Fort Anderson, near the shores of the Arctic Ocean, excepting No. 2, which was found at Great Slave Lake.

LAGOPUS SCOTICUS. Nos. 6, 7, 8, 9 (8545, **A.** 8545, **B.** 8545, **C.** 2407) were sent to me by Mr. J. H. Dunn, who collected them in Scotland.

BONASA UMBELLOIDES. Nos. 11, 12, 13 (5014, **A.** 5014, **B.** 5014, **C.**) were procured at Fort Simpson.

BONASA SYLVESTRIS. Nos. 14, 15 (1331, **A.** 1331, **B.**) were obtained by Mr. A. Newton in Finland.

BONASA SABINEI. No. 16 (6686) was brought from Victoria, Vancouver's Island, and is the only one ever obtained.

BONASA UMBELLUS. Nos. 17, 18, 19, 20, 21 (4312, 4311, 2501, 4772, **D.** 4772, **E.**) Nos. 19, 21 were brought from Canada, 17 from New York, and 18, 20 from Halifax, Nova Scotia.

LAGOPUS RUPESTRIS—var REINHARDTII. Nos. 22 and 23 (3681, 4269) were obtained in Greenland.

LAGOPUS RUPESTRIS. Nos. 24, 25 (7641, 7642) were found near Anderson River.

LAGOPUS RUPESTRIS—var ISLANDICUS. No. 26 (8544) was procured near Vap-na-fjord in Iceland.

LAGOPUS MUTUS. Nos. 27, 28, 29, 30 (8546, **A.** 8546, **B.** 8546, **C.** 8546, **D.**) were taken by Mr. J. H. Dunn in Scotland.

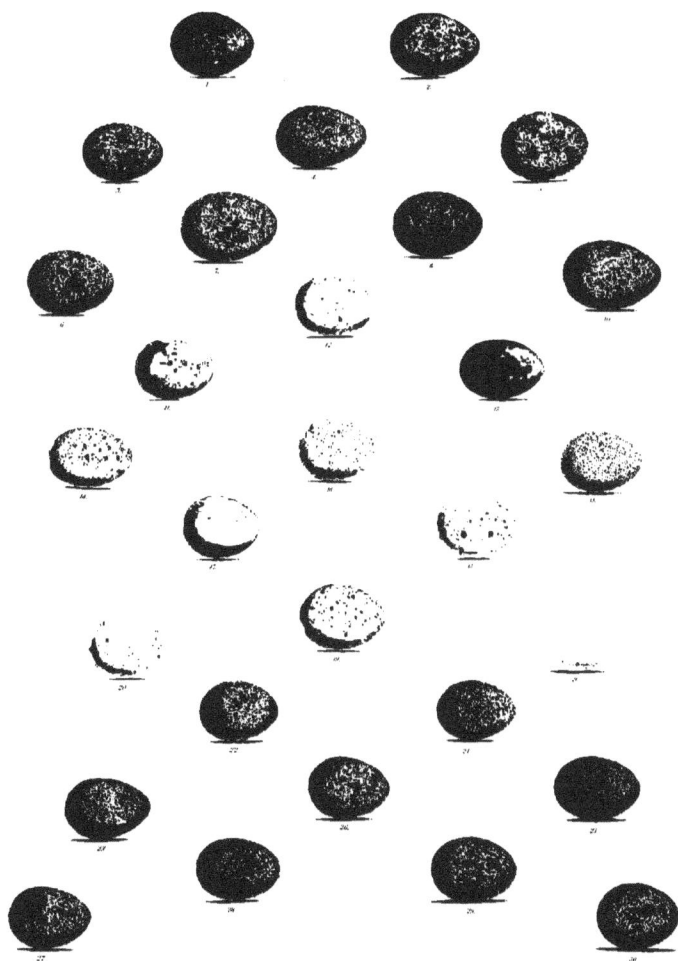